Illustrated Guide to
X-RAY TECHNICS

Illustrated Guide to
X-RAY TECHNICS

John E. Cullinan, R.T., F.A.S.R.T.

Administrative Assistant for Technical Affairs,
Division of Radiology, Albert Einstein Medical Center, Philadelphia

150 Illustrations

J. B. Lippincott Company

Philadelphia · Toronto

Distributed in Great Britain by
Blackwell Scientific Publications,
Oxford and Edinburgh

ISBN 0-397-50284-2

Library of Congress Catalog Card Number 71-165975

Printed in the United States of America

1   3   5   6   4   2

Preface

This textbook is for the student of radiologic technology, whether he or she is a young person at the start of his or her career, a seasoned technologist, a resident in radiology, a practicing radiologist, or a physician who uses radiographic equipment in his specialty. Its aim is to help all these to understand radiographic equipment and accessories better and to use them more effectively. In view of the ever-increasing sophistication of instrumentation and technics, this objective is especially important.

This book makes available to the radiologic technologist, at any stage in his career, information which will enable him to understand the basics of the field and to comprehend technical advances as they occur. Practical information and "rules of thumb" will be stressed. Physical principles, although adequately surveyed, will not be given deep coverage, since this is primarily the province of the physicist and not the technologist.

Most radiologic technologists have their own ideas about technical factors and use them effectively; there is no substitute for experience. The kilovoltage range required for the higher-ratio grid technic would be technically offensive to the user of the lower-ratio grid. Conversely, kilovoltage variations that produce changes in the lower-ratio grid technics would not even be noticed by the advocate of the higher-ratio grid. This book introduces a variety of practical methods. Everyday basic accessories, such as intensifying screens, collimators, grids, Buckys, and timers are explained. The more complex procedures and equipment, including image intensifiers, cineradiography, serial film changers, and video tape recorders, are described. This text avoids large volumes of technical material.

Minor technical error, rather than the malfunctioning of equipment, is the most common enemy of radiographic excellence. The supersophisticated equipment of today tends to overglamorize the work of some: for example, the special-procedure student technologist wants to participate in cerebral angiography before he has mastered the principles of routine radiography of the skull. Hours are spent teaching students to operate the complex equipment needed for angiography—time that could be better spent on problems associated with the technical factors or the routine positioning of conventional radiography. In this book emphasis will be placed on why a specific piece of equipment or an accessory is used with a specific technic, how the combinations work, and what their advantages and limitations are.

During the early years of my practice of radiologic technology, I decided not to attempt to master the specialties which now require independent certification by the American Registry of Radiologic Technology. These include nuclear medicine technology and radiation therapy technology. It is virtually impossible to become expert in this fields and in diagnostic radiology; but there are areas in diagnostic radiology that I feel strongly should be included in the training of all technologists who are interested in a position of responsibility. These areas include, for example:

1. Basic concepts of image detection devices.
2. Principles of collimation, secondary radiation reduction, etc.
3. Newer x-ray tube designs and their usage.
4. Mobile radiography, particularly in the operating room.
5. Body section radiography, which should no longer be thought of as a special procedure.

The intent of the author in this book is to help those engaged in limited practices. It will also be of considerable value to the business-minded administrator, as his technologists learn to get the most out of existing equipment, and to those in positions of authority, charged with the responsibility of choosing new radiographic equipment. The right accessory in combination with an existing installation can result in a rewarding improvement in departmental technic. Often a new piece of equipment is installed, and the quality of the radiographs is disappointing. It is easy to blame the manufacturer, but frequently poor technical quality can be traced to a lack of awareness of the most basic concepts on the part of the manufacturer or the buyer, or both.

This book is primary in nature, and it is as accurate as possible. Wherever possible, technical accomplishments that were arrived at empirically by the author have been substantiated by reference to similar writing of technical experts.

J.E.C.

Foreword

Advances in radiologic technology are continually being made, and in order for the profession to take advantage of this knowledge adequate dissemination of the information is required. A refreshing approach to the presentation of practical aspects of radiographic techniques utilizing currently available equipment, film, screens, grids and other devices has been compiled in a single volume. This is not another manual of radiographic positioning and x-ray technics oriented towards "push-button" radiography.

The modern x-ray technologist, whether in a one man office or a large multi-specialty hospital department, should know what is available in radiographic tools, how and when to use them to best advantage and what to anticipate as an end product for optimum diagnostic purposes. A book that provides such information is especially appropriate at a time when programs for the delivery of medical care utilizing the services of trained medical assistants are being established in many areas.

The contents of the book include lucid, comprehensible descriptions of modern so-phisticated equipment, discussions of special technics and their appropriate applications, nuggets of information that can spell success or failure in critical situations and in addition, a review of basic principles that need emphasis and are frequently ignored in every day practice. Throughout the text the descriptive material is highlighted by excellent illustrations.

For many years Mr. Cullinan has taught and lectured to x-ray technologists and radiology residents on the subjects which he has elected to present and with which he is fully conversant. I consider it particularly fortunate to have him associated with our department for all of us frequently seek his advice and help in technical matters. Congratulations to the author are in order for his initiative and recognition of the need for a readily available technical manual for the radiology profession.

Harold J. Isard, m.d.,
Chairman, Division of Radiology
Albert Einstein Medical Center
Northern Division
Philadelphia

Acknowledgments

I appreciate the encouragement of my friends over the past 25 years, especially my first instructors and the late James Loftus, a special friend.

I would like particularly to thank the technical and professional staffs of the Department of Radiology of The Albert Einstein Medical Center for valuable advice and encouragement, especially Warren Becker, M.D., who is as good a radiologic technologist as he is a radiologist. My most sincere appreciation to Harold J. Isard, M.D., Chairman of the Division of Radiology of The Albert Einstein Medical Center for his friendship and his genuine interest in my future.

I want to thank Mr. John B. Cahoon, R.T., F.A.S.R.T., and Mr. Leonard Stanton, M.S., for their continuing inspiration.

Sincere thanks to my associates in the commercial aspects of radiography: Mr. Arnold Auger, Mr. Jack Berry, Mr. Walter Butterworth, Mr. Terry Eastman, R.T., Mr. Harold Smythe, Mr. Robert Trinkle and Mr. Charles Worrilow. Their advice is highly respected.

A special mention is due the Audio-Visual Department of The Albert Einstein Medical Center, under the direction of Mr. Ken Goodman, for the making of the photos and illustrations used in this textbook.

The professional staff of the J. B. Lippincott Company has been very helpful. One member of that organization must be singled out—Mr. George F. Stickley, Project Editor of the Medical Books Department—for his help and above all his faith in my ability.

A special remembrance is noted to my parents; to my mother who would have been pleased with her son's first book and to my father who, I hope, will enjoy my efforts.

To my children, Jeanne, Patty, Diane and Teresa, I offer a word of thanks for giving up their father for a good portion of the past two and a half years.

And finally, to my wife, Angie, gratitude for typing, editing, reviewing, and giving technical advice. She did everything that a good friend could do short of actually writing this textbook.

JOHN E. CULLINAN

Contents

To Angie, my wife, who is also my favorite R.T. (A.R.R.T.), A.S.R.T.

1. Basic Image Detector Principles

A review of the variety of image-detecting devices available is presented in this chapter (Fig. 1-1). Types of film, film speeds, the cassette, intensifying screens, film-screen combinations, and fluoroscopic screens will be considered in relation to their separate or combined effectiveness. Their merits, difficulties, and potential will be evaluated. Their care will be described.

It may seem incredible to the practicing technologist that after spending 2 years in an intensive type of photographic training he should be restricted to a single-speed film and a single-speed pair of intensifying screens. Anyone who has used a simple camera becomes rather quickly aware of the multitude of technical combinations available to the amateur photographer. Yet some radiologic technologists restrict themselves or are restricted to a 1-speed film-screen combination. It is true that the trained technologist does not have at his disposal the variety of technical possibilities of the trained photographer, but he does have at his command 3 or 4 types of x-ray film and/or intensifying screens which, when used for radiographic work, can result in a dozen or more combinations of speed, latitude, and detail.

A type of medical radiographic film known as nonscreen film is generally used for direct-exposure radiography of the extremities or of soft-tissue structures. Conventional radiographic film generally used in combination with intensifying screens is available in several speeds, the most popular being *average* or *normal. High-speed* radiographic films which are approximately 50 per cent faster than average-speed film are avail-able. Some manufacturers market a type of radiographic film which could be classified as *slow-speed* film.

Intensifying screens are available in several speeds, *slow, medium, fast,* or *very fast,* although there is some variation in screens of similar speeds due to different methods of manufacture. If all intensifying screens and all radiographic films could be rated with a guaranteed intensification factor, and all would respond in the same manner over the entire range of diagnostic kilovoltage, there still would be a dozen or more image-detection combinations. But such an assumption is invalid, for films and screens do vary according to the specifications of the manufacturer and can respond differently in intensification over the entire diagnostic kilovoltage range.

The following example will illustrate the wide variety of technical exposure combinations possible. If a moderately slow film with slow intensifying screens were used in an exposure range of 75 kVp at a hypothetical 300 mAs value (100 mA × 3 seconds), a duplication in radiographic intensity equal to the initial exposure but utilizing a high-speed film with very-fast-speed screens would require the initial kilovoltage range of 75 kVp but only 100 mA at $3/_{10}$ of a second, for a total of 30 mAs. There is here a 10-fold difference between the intensification factor with the slow screen and the slow film and that with the fast film and the very fast screen. Unfortunately, this gain relationship is rarely linear, but this comparison gives an idea of the possibilities available. The key word here is *available,* for the radiographic products stressed in this chapter are easily

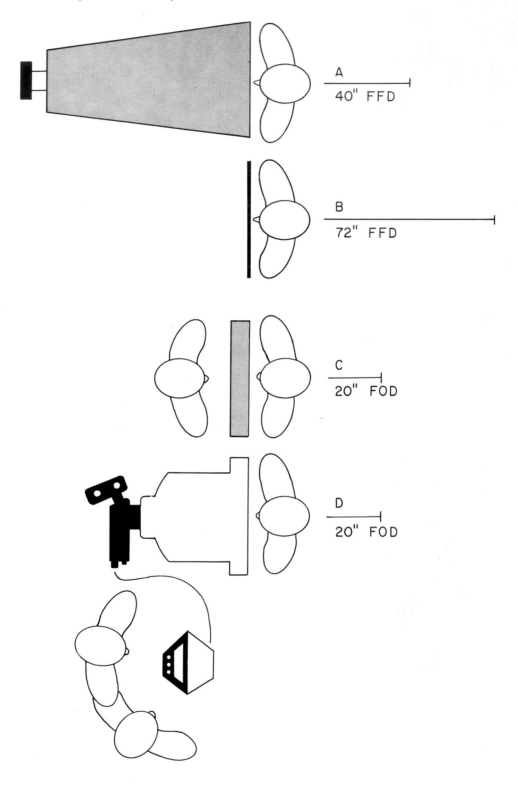

A
40" FFD

B
72" FFD

C
20" FOD

D
20" FOD

secured from local commercial representatives.

Although much of the material presented in this chapter may be of little or no value in some clinical facilities, an awareness of these film-screen combinations enhances intelligent use of existing equipment if and when the need for technical variations arises. It is important to know the information in this chapter even if it is never used directly. Knowledge in itself is a reward, and the technologist who has improved his technical approach to routine problems is a better technologist.

MEDICAL SCREEN X-RAY FILM

Medical screen x-ray film is used in combination with intensifying screens for medical radiography. This film is composed of an inert plastic base, generally polyester or cellulose acetate, both sides of which are coated with an emulsion of silver halide crystals, usually silver bromide, suspended in a gelatin matrix. An adhesive assures the binding of the silver crystals to the base of the film. The silver crystals are extremely sensitive to all forms of light, x-rays, and gamma rays. There is a fairly constant rela-

---

Fig. 1-1. Basic image detection devices.

(A) *Photofluorographic unit.* A fluoroscopic screen is housed within a lightproof hood. A camera is attached for the making of a photographic image of the fluorescing full-size screen. 70 mm strip film or 4″ x 4″ cut film is utilized to record the fluoroscopic image. The patient is positioned as for conventional chest radiography. The shortened focus-film distance, approximately 40 inches, is used in conjunction with an automatic timing system. Elaborate light gain systems using mirror optical systems are available. Speed, efficiency, and economy are associated with this type of image detection device.

(B) *Conventional radiography.* A radiographic cassette utilizing a pair of intensifying screens is used in conjunction with medical screen x-ray film. A 72 inch focus-film distance is used for *teleoroentgenography* to minimize geometric enlargement and distortion. A 40 inch focus-film distance is suggested for all conventional cranial, trunk, or extremity radiography. When activated by x-radiation, the intensifying screens expose the dual emulsion radiographic film.

(C) *Direct fluoroscopic viewing.* A fluoroscopic screen is substituted for the photofluorographic unit or conventional cassette, enabling the fluoroscopist to view anatomic details as well as physiologic activities. The shortened focal object distance of approximately 20 inches is used with this system. The focus-film distance is determined by the placement of the fluoroscopic screen. The smaller the patient, the closer is the screen,

with a resulting lowering of the focus-film distance. The fluoroscope is used to observe the dynamics of the internal organs, and it is definitely much more economical to operate than conventional filming. Standard fluoroscopic screen viewing requires dark adaptation of the eyes to visualize the dimly lit image in the totally darkened radiographic room.

(D) *Image intensification fluoroscopy.* The advantages of fluoroscopy are multiplied by the utilization of an electronically enhanced fluoroscope, commonly known as an image intensifier. As with conventional fluoroscopy, a shortened focal object distance of 20 inches is utilized with the focus-film distance, being determined by the size or position of the patient. A complex electronic device amplifies the dimly illuminated fluoroscopic image over 3,000 times. This permits viewing of the image without complete room darkness. Mirror optical systems connected to the image intensifier can be used by the fluoroscopist for direct viewing, and a television system can be added to the intensifier, permitting viewing of the fluoroscopy by many individuals. Television monitors can be utilized in remote areas for teaching purposes and video tape recordings can be made of the fluoroscopic image. The high-light yield of the intensification system has made cineradiography of x-ray movies possible. The image intensification system permits instant viewing by many as well as permanent recording of the image.

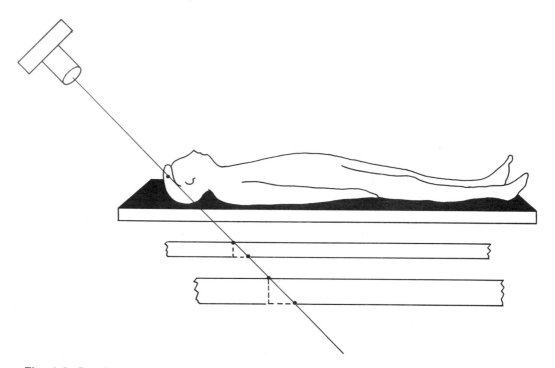

Fig. 1-2. Parallax effect. Note the lessening of the parallax effect by the use of a radiographic film with a thinner base. When a tube-angle technic is used, the separation of the images on dual emulsion film is minimized by the use of a thinner base.

tionship between the speed and the graininess of the film: the larger the grain size of the silver crystals, the faster is the film. The film base acts as a support for the emulsion. An extremely thin base, generally tinted a light blue, is necessary to avoid parallax errors in the radiographic image when the film is viewed from an angle rather than from a direct frontal position.[30] A thinner base is necessary to guarantee the sharpness of the superimposed images when an angled tube technic such as the Chamberlain-Towne projection is used (Fig. 1-2). A very thin supercoating is added to the surface of the film to prevent abrasion to the emulsion (Fig. 1-3).

DIRECT-EXPOSURE MEDICAL X-RAY FILM

Direct-exposure medical x-ray film (or nonscreen film) is slightly thicker and has a larger grain size of silver crystals than screen-type medical radiographic film. A considerably larger number of layers of sensitive silver crystals increases the image-resolution potential of direct-exposure film (Fig. 1-3). The use of a thicker emulsion with increased silver content is advantageous for direct-exposure technics because crystal exposure is affected only by the penetration of the x-ray beam, whereas in intensifying-screen-medical film technics more than 99 per cent of the blackening of the film is caused by the blue-violet light of the activated intensifying screens. Nonscreen film should not be used in combination with an intensifying screen because its primary sensitivity is to x-radiation and not to blue-violet light. The light given off by the fluorescing calcium tungstate crystals of an intensifying screen would affect only the silver crystals on the surface or adjacent to the surface of the thicker nonscreen emulsion. Therefore, to be fully exposed, the thicker layer of silver crystals found in nonscreen film requires a form of energy penetration such as x-radiation.

Nonscreen medical film is used with a lightproof holder, such as a cardboard, paper, or flexible vinyl envelope. These radiolucent holders are light-protective to the film. Most cardboard holders contain thin sheets of lead foil on the undersurfaces to minimize fogging of the x-ray film by *backscatter*. While nonscreen film can be loaded into cardboard holders by the user, most manufacturers prepack this direct-exposure film in lightproof holders (Fig. 1-4). Prepacked direct-exposure holders with lead foil backing are not available, and if backscatter is of concern to the technologist, a sheet of lead rubber

can be placed under the holder during an exposure.

The popularity of nonscreen medical film dropped in the late 1950's as the automatic film processors with their predetermined developing and fixing cycles came into common use. The thickness of the emulsion of the direct-exposure film generally necessitates a relatively long developing time to blacken satisfactorily the exposed silver crystals and a long fixing time to clear the unexposed layers of silver crystals. Unfortunately, most available nonscreen film still must be hand-processed and requires the use of conventional time and temperature technics. Using nonscreen film becomes not a matter of personal preference but rather a matter of logistics, for a busy department

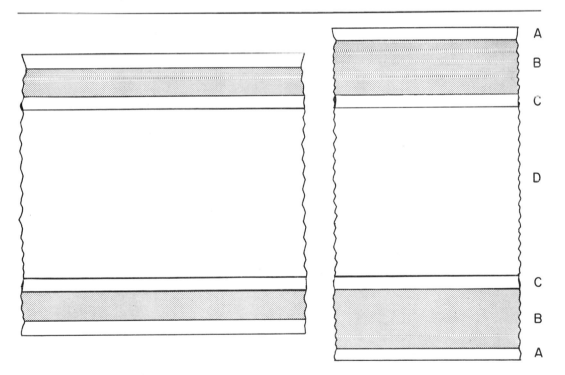

MEDICAL SCREEN FILM   DIRECT−EXPOSURE FILM

Fig. 1-3. Basic medical radiographic films. Both medical screen and direct exposure (nonscreen) radiographic films are made with a base (D) of a similar thickness. The emulsion of the screen film is sensitive primarily to the blue-violet light of a fluorescing intensifying screen. The thicker emulsion of the direct exposure film is sensitive primarily to x-radiation.

Both films use a subcoating (C) of a thin adhesive type of material to bind the sensitive emulsion (B) to the film base (D). A protective coating (A) covers the emulsion.

TUBE SIDE | THIS SIDE UP

Fig. 1-4. Direct exposure (nonscreen) film holders. Commercial film holder, marked TUBE SIDE—THIS SIDE UP, is used for direct-exposure technics. Film is loaded into this holder, which contains a thin sheet of lead foil on its undersurface to minimize fogging of the x-ray film by *backscatter.* A prepacked holder is shown with the corner cut away to demonstrate the film and its protective paper covering.

frequently cannot afford the work-load snarl created by 2 separate types of processing facilities.

A common technical variation for circumventing the hand-processing of nonscreen film is the use of medical screen x-ray film in a nonscreen holder. Since medical screen film is more sensitive to blue-violet light, it is more inefficient than nonscreen film for direct-exposure technics. The sensitivity of nonscreen film to x-radiation can be 3 to 4 times that of conventional medical screen film. A distinct advantage of the use of medi-

cal screen film for the direct-exposure technic is that an automatic processor can be used to develop this film, thus expediting patient flow.

A compromise between the direct-exposure technic and a conventional screen exposure is the use of a cassette with slow-speed intensifying screens. When slow-speed screens for radiography of the extremities are used, a 50 per cent increase over the mAs values of medium screen technics is generally adequate for comparable film blackening. Because of the latitude of the slow-speed screens, technical errors are not as critical as with conventional screen technics. Routine extremity radiography with slow- or medium-speed screens is quite acceptable although image quality suffers when compared to that in the direct-exposure technic.

RADIOGRAPHIC FILM SPEEDS

Several types of radiographic film, generally referred to as regular-speed or high-speed x-ray film, are available. High-speed films are more susceptible to background radiation, and some of these films age poorly; that is, they have a short shelf-life. It is important to order only film that will be actually used rather quickly. Although some high-speed films tend to look grainy, many of them have very satisfactory contrast levels, and their use has become a matter of personal choice. The use of these faster films should be determined by comparison tests made under actual working conditions, for the proper use of technical skills in combination with proper technic and accessories can improve or destroy the look of a film. A capable technologist can make almost any medical radiographic film look or perform satisfactorily, regardless of its inherent speed or contrast. Too often, the misuse of a film product by a technologist—for any number of reasons, such as poor screen contact, poor primary beam limitation, the use of a lower ratio grid with a high kilovoltage technic, improper safe lighting in the darkroom—will result in rejection of a specific film.

Often one hears that a specific film is 2 kVp faster or 2 kVp slower than another film, or that a specific film has an inherent base fog of 2 or 3 kVp. It is quite interesting to travel the different sections of the country and question technologists about these seeming variations. Film that is reputed to be 2 kVp faster in one area can be said to be 2 kVp slower in another. The base fog of one film that is unacceptable in one locality can be thought of as crystal-clear in another. Most opinions about minor film variations are just that—opinions—and have little or no basis in fact. Classic experiments in kVp variations by Fuchs and Cahoon[2,7] reinforce this statement. Most of their early experiments were made without the benefit of the higher ratio grid and efficient beam-limiting devices which improve present-day latitude capability.

Minor variations in film speed are more imagined than real. It is important to stress that processor temperatures affect greatly the so-called "speed" of radiographic film. Minor temperature variations in the developer section of the automatic processor can produce major density changes on the finished radiograph (Fig. 1-5).

There is an increasing trend towards high-speed film and/or fast-speed screens in conjunction with x-ray procedures that require a reduction in radiation dosage to the patient. These combinations permit a shortened exposure time to overcome object motion or to augment equipment of limited output, such as a bedside radiographic unit. Some of the more common uses of high-speed film or fast-speed screens or their combination include portable radiography, operating room studies in which it is difficult to control the patient's breathing, radiography of the pregnant abdomen, in which both radiation dosage and shorter exposure time must be considered, pediatric technics, and fluoroscopic spot film studies.

Several advantages attend the use of high-speed films or fast-speed screens during fluoroscopic spot filming. First, dosage is lower to the patient. This is particularly important because of the shortened focal-object distance generally used in fluoroscopy, approximately 18 inches (Fig. 1-1). Second, the significant shortening of exposure times helps to overcome normal or hyperactive organ motion. Third, spot film radiographs are usually made with the central ray perpendicular to the film; therefore, the grain of the fast screens or fast film is not exaggerated by tube angulation. Fourth, the resultant gain of light intensity is accomplished by shortened exposure times, a feature which helps to extend the life of the x-ray tube. The total

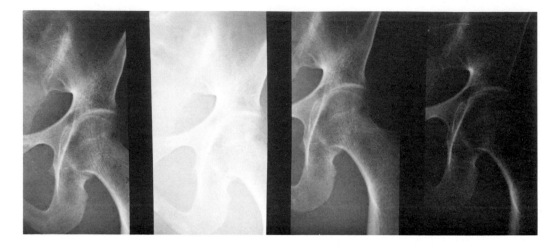

Fig. 1-5. Density variations due to automatic processor difficulties. In a large hospital where many automatic processors are available, particularly at remote or satellite locations, it is possible to have severe differences in radiographic densities due to mechanical or chemical problems with the automatic film processing systems.

Four radiographs of a pelvic phantom were made, utilizing the same radiographic unit, the same cassette, the same technical factors, and the same collimation method. The only variation in this experiment was the processing of individual films in different processors. The 4 radiographs shown above were processed in 4 separate processors, all within 1 hour, in a large metropolitan hospital. Two of the radiographs seem of average density, one is quite dark, and the other unreasonably light. These variations can be due to any number of reasons and may go unnoticed unless direct comparison films are obtained. Quality control is a *must* before attempting new radiographic technics, particularly in remote areas.

---

heat rating capability as well as the individual exposure tolerances of the x-ray tube can be held well within the specifications of the manufacturer.

PROCESSING DIFFICULTIES WITH FASTER-SPEED FILMS

With the development of the automatic processor, processing difficulties diminished, and film quality vastly improved, if for no other reason than that the automatic processor eliminates the possibility of sight development. A preset, established develop-fix-wash-drying time has significantly improved quality control. A serious difficulty still plaguing the technologist in the darkroom is the lack of concern for safelight fogging. Unexposed film can be handled for a relatively long time—for example, 1 minute—whereas exposed film is considerably more sensitive to safelight fogging and must be handled quickly and carefully. Some of the faster speeds of medical x-ray film are extremely sensitive and require the use of a $7\frac{1}{2}$-watt bulb in an appropriate safelight, whereas conventional radiographic film can tolerate the standard 15-watt bulb.

Medical x-ray film of a slightly slower speed than regular-speed x-ray film is available for use with intensifying screens. Exposure latitude is considerably improved due to the increase in the margin of error effect of the slower emulsion. In actuality these films are not slower than average-speed films, but they do have a lower fog level because of their lack of sensitivity, and therefore they require a slight increase in exposure.

Medical x-ray film is available with a single coating of emulsion. This film, which has a blue-tinted transparent safety base with an antihalation backing to absorb the back-scatter of light within the cassette, improves image quality. It also helps to avoid parallax difficulties in body section radiography or in tube-angle technics.

CASSETTE

The major function of the cassette is to protect the sensitized emulsion of radiographic film from light. Cassettes are generally quite sturdy, with a radiolucent front, such as bakelite or magnesium. Intensifying screens are mounted within the cassette on the front and back sections, and a locking mechanism is provided with some type of light sealant, such as oil-free felt, to provide for maximum screen-film contact. A thin sheet of 0.005 inches of tin-lead foil is mounted on the back of the cassette underneath the posterior intensifying screen (Fig. 1-6) to absorb backscatter during exposure. A typical cassette requires a pair of intensifying screens; the speed is predetermined by departmental policy.

Some producers of intensifying screens make the front intensifying screen thinner and the back intensifying screen thicker. Theoretically, the thinner screen is used so that remnant radiation exiting from the patient will not be severely attenuated by the cassette front or the thicker anterior screen. The thickness of the lead on the inner surface of the back of the cassette varies with the conventional cassette and the so-called photo-timing cassette. Phototiming cassettes have a thinner sheet of antimony-lead foil (0.0025 inch). This thinner lead foil is required in using a phototiming mechanism, which will be described in Chapter 5.

It is important to use an identification system for cassettes and intensifying screen pairs. For example, the use of the letter "P" prior to the number in the cassette for par or medium-speed screens is recommended. Lead marking numbers are placed in the cassette by the manufacturer, and the back of the cassette is labeled correspondingly.

Knowing the speed of the screens in use for a particular study is of great help to the quality-control technologist when he is critiquing radiographs. It is very easy for a busy technologist or student to use inadvertently under pressure the wrong type of intensifying screens.

INTENSIFYING SCREENS

The use of an intensifying screen in contact with radiographic film is not a new technic. On February 7, 1896, Professor Pupin of Columbia University in association with Thomas A. Edison utilized the first screen-film combination to produce a radiograph of a hand ridden with buckshot.[1] Intensifying screens were evaluated in Italy, France, England, and America by physicists during the first 3 months of 1896. By the end of that year, Levy in Berlin was using radiographic film, coated with emulsion on both sides, placed between 2 intensifying screens.[5]

Yet for a variety of reasons, such as poor screen life, screen grain, etc., intensifying screens coupled with radiographic film were not popular during the early years of radiography. Grain size is still a minor problem, but severe graininess was a serious problem to the pioneer specialists in the late 1890's because of the large crystal size then in use. Graininess was not the only difficulty encountered by these roentgen innovators. The early crystals lacked uniformity of size and varied in fluorescence so that there was frequently an objectionable lag or afterglow (phosphorescence) produced by the inherent impurities. Today, because of improved manufacturing methods, radiologic technologists are relieved of these problems.

Intensifying screens glow blue-violet when activated by radiation. The crystals of these screens are usually composed of calcium tungstate, a natural mineral. Fortunately, it is no longer necessary to depend on nature as the source of calcium tungstate crystals, for it is now possible to synthesize these luminescent crystals. Today all conventional calcium tungstate intensifying screens are

Fig. 1-6. Cassette components. A sectioned view of a typical radiographic cassette includes: (1) a bakelite cover, (2) a metal base for the cover, usually magnesium, (3) the front intensifying screen, (4) a medical screen radiographic film, (5) the back intensifying screen, (6) oil-free felt lining to act as a light sealant as well as to aid screen-film contact, and (7) a thin sheet of lead foil to absorb *backscatter*.

made from manufactured crystals rather than the natural mineral. Fluorescent crystals are applied to a base of cardboard or plastic. They are mixed with a binder and added in multiple thin, even coatings. The front intensifying screen, as we have said, can be somewhat thinner than the back intensifying screen to lessen radiation absorption. A protective surface coating is applied to the finished screen. Although most screens are made of calcium tungstate, some fast-speed intensifying screens, utilizing a barium-lead sulphate crystal as a fluorescent source, have been available since 1948.

Intensifying Screen Crystal Size

Much has been said about the crystal or *grain* size of the fluorescent chemicals used and their effect on the speed or gain of an intensifying screen. It is generally reported that these crystals are in the range of 5 microns in size. Although the size of the intensifying screen crystals influences the speed of the intensifying screen (the larger the crystal size, the faster is the screen), it is not the most significant factor in determining the gain of an intensifying screen. The most important factor in the manufacture of the faster-speed screen is the thickness of the layers of fluorescent crystals: generally, the thicker the layer, the faster is the intensifying screen speed. Handee in a comparison of crystal size and layer thickness reports a crystal size of 4 microns and a layer thickness of 50 microns for the average detail (slow) screen. A typical fast-speed screen has a crystal size of 8 microns and a layer thickness of 300 microns.[12]

The thicker the layer of intensifying crystals, the poorer is the detail of the finished radiograph. It would be ideal if intensifying screens could be made with as thin a layer of fine crystals as possible. However, this is not practical, because thicker layers of large-size crystals are necessary for several reasons: First, they permit the use of technics with significantly lower radiation dosage to the patient; second, they reduce tube loading, thereby contributing to the life of the radiographic equipment; third, they increase the sharpness of the radiographic image by lessening the chance of voluntary or involuntary patient motion.

Types and Speeds of Intensifying Screens

For the sake of simplicity, screens will be referred to as being of a *slow* speed, *medium* speed, *fast* speed, and *very fast* speed. Intensifying screen speed is usually said to be affected by the size of the fluorescent crystals. For example, small crystals are used for slow-speed screens, medium-sized crystals are used for medium-speed screens, and larger crystals are used for fast-speed screens.

Some of the more common terms used for the slow-speed screens are *detail, high resolu-*

tion, and *ultra detail.* For the medium-speed screen the terms include *general, average, mid-speed,* or *par.* Fast-speed screens are termed *high, fast,* or *very fast.* Many other terms besides those listed above can be used to describe screens of different speeds.

We are told that when activated by radiation, the smaller intensifying crystals give off less light, and the larger crystals proportionately give off more light (Fig. 1-7). This is true, and yet it is important to stress that the thicker layers of crystals have *more* effect on the gain of the screen than do the size of the crystals. The technologist must be concerned with the damaging effect on sharpness of the radiographic image caused by the reflection of light from the underlying multiple layers of crystals of the faster intensifying screens. It would be ideal if a way of collimating this overlap of light were available. This focusing of light would help to eliminate the overlap of light from individual crystals as well as the damaging side reflection from successive underlying layers of the luminescing chemicals. This effect is somewhat accomplished in the slow-speed screens by employing a light-absorbing dye (such as yellow, beige, or pink). The dye is added to the screen to trap or absorb unfocused, underlying, or reflected light. Light rays travel almost directly from the crystals to the film without significant deflection, improving screen definition but significantly reducing screen speed. This problem can be complicated by crossover of light from the front to back intensifying screens due to the clear, light-blue base of medical radiographic film. It would be helpful if this base could be colored and therefore be more opaque to light. But a colored base would create a problem from the standpoint of interpretation, for it would be visually annoying. Another way to achieve increased screen definition would be to reduce phosphor size (by using smaller crystals) as well as to reduce the thickness of the phosphor layers. The screen surface staining process is the most common method used to produce a slow-speed or detail screen.

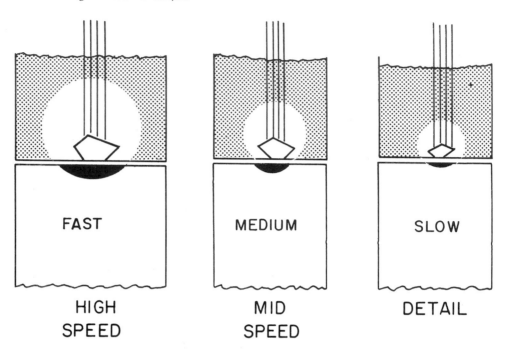

Fig. 1-7. Intensifying screen crystal size comparison. Ideally, crystal size should diminish as screen speed lessens. In general, the larger the crystal size, the more light is emitted by the crystal when exposed to x-radiation. In practice, layer thickness of the crystals, as well as actual crystal size, influences the speed of an intensifying screen.

Intensifying Screen-film Contact

Every effort should be made to guarantee maximum intensifying screen-film contact, for poor screen contact significantly lowers contrast and sharpness in the finished radiograph. Screen contact can be tested by making a radiograph of a $\frac{1}{4}$ inch wire mesh placed in direct contact with the cassette. A low kilovoltage exposure is made and the radiograph evaluated for a loss of image sharpness.

An interesting and effective method of evaluating the wire mesh radiograph by using red accommodation glasses to block out effectively the excessive light of the x-ray view box was reported by Horner. By filtering the large areas of light seen from fully illuminated view boxes, areas of poor screen contact which would ordinarily be missed can be easily localized. Horner also suggests that by standing in an oblique position in relation to the view box one may look at the radiograph at an angle and so avoid the excessive amount of light given off by the ordinary view box.[13] If red accommodation glasses are a thing of the past in your department, viewing the screen contact radiographs from a 10 or 12 foot distance will produce much the same effect. Since areas of poor screen contact can be defined by their cloud-like appearance, a gray, hazy increase in density can easily be seen from this increased distance.

Poor screen contact is quite common where there are severe temperature and humidity changes, and it is felt that the material used as a cassette front may be at fault. Magnesium, when substituted for bakelite, virtually eliminates warping due to high temperatures or changes in humidity, thereby preventing, or at least lessening, disturbances in screen-film contact.

It is imperative that absolute contact between the screens and film be maintained, because using an intensifying screen produces a type of "contact" print on the radiographic film. As previously mentioned, light arising from the lowest level of the layers of crystals in an intensifying screen produces significant image unsharpness. It is therefore doubly important that contact be maintained so that the light from the intensifying crystals is received by the film before it has a chance to diffuse. Experiencing bad screen contact in the smaller cassette sizes is unusual; the larger 14" × 17" cassettes are more prone to this difficulty.

Screen Resolution

In photographic physics it is customary to measure the resolution of a photographic material by focusing upon it a test pattern consisting of a series of black lines on a white background, where the widths of the lines are equal to those of the spaces between them. Resolution is then expressed in terms of the maximum number of lines per millimeter which the photographic material is capable of recording. Quimby reports that in evaluating intensifying screen resolution, values in excess of 20 lines per millimeter have been recorded.[9]

Some of the questions that must be asked prior to accepting resolution tests are: (1) What was the quality of the x-ray beam used for this test? (2) What density level was exhibited on the film? (3) What focal film distance and focal spot size were used? (4) What were the processing conditions? (5) What were the viewing conditions? (6) Was the human eye used to make this test?

Listed below are the published lines per millimeter of the resolution capabilities of the intensifying screens manufactured by the United States Radium Corporation:

U.D.(*slow*)	14 lines per mm
T-2(*medium*)	12 lines per mm
TF-2(*fast*)	10 lines per mm
STF-2(*very fast*)	8 lines per mm

Comparison Between Direct-exposure and Screen Technics

Before making an attempt to compare the look of a screen film used with intensifying screens and the look of direct-exposure film, it should be stated that such a comparison is not really valid, because we would be comparing 2 different image-detection methods. The direct-exposure film has a flat appearance and is almost devoid of contrast, whereas the screen-film exposure exhibits extreme contrast. It is possible but not practicable to compare their relative speeds under ideal conditions—that is, with a perfectly matched set of view boxes having equally balanced lights, a perfect processing system, and a completely objective viewer. The decided differences in contrast levels between the 2 types of film constitute the primary difficulty in making a meaningful comparison.

A distinct advantage of the direct-exposure technic is that it offers a wide range of exposure latitude. Not only is there a marked reduction in latitude when intensifying screens and film are used, but this latitude decreases proportionately and sometimes drastically as the speed of the intensifying screens and films increases. The ideal combination of contrast and latitude is difficult to find since one of these factors cannot be changed without affecting the other. As contrast increases, latitude decreases; and as contrast decreases, latitude increases.

Stanton defines *latitude* as the ability of an examination to demonstrate satisfactorily on the same roentgenogram objects of markedly different roentgen absorption.[23] For example, chest radiographs should demonstrate not only ribs, spine, hilar soft tissues, and the lung fields, but also other details.

The speed-ratio differences between intensifying-screen-film combinations and film in a direct-exposure holder has been a source of considerable disagreement among experts. Many technologists still erroneously refer to any type or speed of film in a nonscreen holder as a nonscreen film. Mistaken speed comparisons are frequently made between 2 films because of a lack of awareness of the speed differences. For instance, a "sandwich" radiograph of a foot was made in which direct-exposure nonscreen film was used in the same cardboard holder with medical screen x-ray film. The density differences are obvious (Fig. 1-8). Several radiographs are shown to demonstrate density differences. These films were taken with increasing milliampere-second values and vary in density in a step-wedge fashion (Fig. 1-9).

It is most unlikely that any group of radiologic technologists or radiologists would agree on a proper density level. It is not the purpose of the author in this textbook to suggest optimum density technics; yet it is interesting to discuss material of this nature with prac-

ticing technologists. Instead of documenting educated guesses, however, a review of the technical literature will be presented to illustrate conflicting opinions and why they exist.

Stanton states that fluorescing intensifying screens can vary from 10 to 40 in film blackening effect as compared with even the fastest direct-exposure film. All but 3 per cent or less of the final image is due to the light of the intensifying screen.[24] In *Fundamentals of Radiography,* a gain of at least 15 times is quoted in one section of the booklet, but a

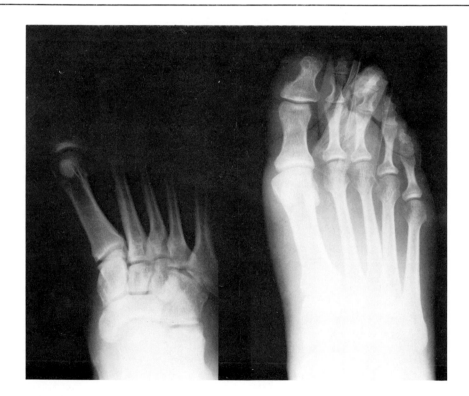

Fig. 1-8. Speed differences between medical screen film and direct-exposure nonscreen film. A cardboard holder was loaded with an average-speed medical radiographic film and a typical nonscreen exposure film. An exposure of the foot was made using this "sandwich" film technic.

Note the differences between the screen and the nonscreen film.

The darker radiograph or nonscreen film exhibits good detail in the thicker portions of

the foot with almost complete obliteration by blackening of the distal portions of the foot. The lighter or screen radiograph exhibits good technical detail in the thinner portions of the foot. The bones of the tarsal area are barely visible. Both of these radiographs were made simultaneously in one direct exposure holder, the only variation being the speed of the films.

Nonscreen film is significantly faster than screen film when used for direct-exposure technics.

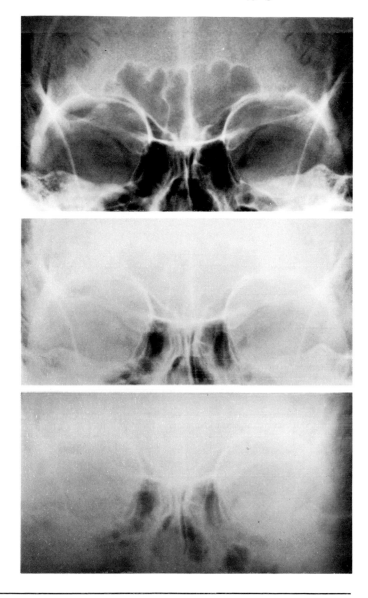

Fig. 1-9. Major differences in density. Three films were made of the skull in a modified Caldwell position. The same cassette and kilovoltage level were used for all 3 views. The top frame was exposed at 100 mAs, the center frame at 50 mAs, and the bottom frame at 25 mAs. It is difficult to believe that there is a 400 percent difference in density between the top and bottom frames. A tremendous amount of technical latitude is available to the present-day technologist.

more practical estimate of a gain of 25 is suggested in a later section of the booklet.[28] Fuchs, in *Principles of Radiographic Exposure and Processing*, suggests a gain factor of 20 to 1 or 30 to 1, depending on the type of intensifying screen used.[8] In a booklet entitled *Sensitometric Properties of X-ray Film* (published by the Eastman Kodak Co.), a wide variation in speed from 15 to 50× under practical conditions is given.[21] A current technical booklet, *The Care and Use of Cronex XTRA LIFE Intensifying Screens* (published by E. I. du

Pont de Nemours & Co.), states that an average gain value of 40 to 50× is achieved in moderate kilovoltage range.[27] The intensification factor of any given screen may vary, depending on the film or kilovoltage range used. Selman states that at 40 kVp the speed factor with medium-speed screens is about 30, whereas at 80 kVp it increases to about 75.[20] Cahoon states that 99 per cent of the total exposure is effected by "printing the visible image onto the x-ray film." X-rays are responsible for only about 1 per cent of the total exposure.[3] These conflicting but honest differences of opinion are not documented to confuse the reader, for each statement is true after some qualification.

How can we intelligently discuss this gain factor when experts seemingly disagree? A good practical rule of thumb would be a gain of $25\times$ in intensification for medium-speed screens. Since fast-speed intensifying screens are reputedly twice as fast as medium-speed screens, a gain factor of 50 would be an acceptable estimate in using fast-speed screens with conventional screen film as compared with conventional screen film in a nonscreen holder. Slow-, average-, or fast-speed films can be used in a direct-exposure holder as a substitute for nonscreen film. Imagine using any of the screen films in conjunction with slow-, medium-, fast-, or extra fast-speed intensifying screens and further compounding these technical variations by a density difference in cassette fronts. All of the available variations are not listed. A report from the U.S. Department of Health, Education, and Welfare states that the number of available radiographic films and fluorescent screens is such that they can be combined to make over 200 different photoreceptors.[22]

Relationship of Film Grain Size to Intensifying Screen Crystal Size

Most intensifying screens have an actual crystal size in the 5 micron range; yet we constantly hear about the graininess of x-ray film rather than the graininess of intensifying screens. Photographic associates seem appalled at the grain size of typical medical x-ray film. It must be admitted that by photographic standards conventional radiographic film is of poor quality, for the average grain size of radiographic film is in the $1\frac{1}{2}$ micron range (Fig. 1-10). Film is not the limiting factor in resolution capability. *The weak link in the resolution chain is the crystal size of the intensifying screen.* Remembering the "contact" print effect of the intensifying-screen-film combination, one notes that the basic image detector, which is the intensifying screen, is approximately 3 times more grainy than the secondary image detector, the radiographic film.

Optimum Film-screen Combinations

Since the use of film in combination with intensifying screens causes some unsharpness, there is no perfect combination of intensifying screens and radiographic film. The use of faster film and screen combinations can provoke a scintillation effect known as *quantum mottle.* Mottle appears as a granularity or spotty nonuniformity of the image. Mottle can generally be attributed to the fact that the x-ray beam is composed of individual photons, and when faster detection combinations are used, the number of x-ray quanta impressed upon the recording system is decreased. When a slower film is used—even

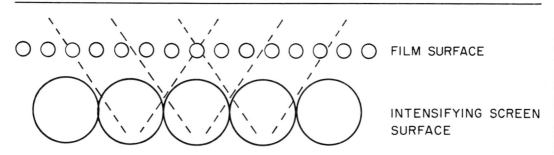

FILM SURFACE

INTENSIFYING SCREEN SURFACE

Fig. 1-10. The relationship of film grain to intensifying screen crystal size. Most intensifying screen crystals are in the 4 to 5 micron range, where as the typical grain size of radiographic film is about $1\frac{1}{2}$ microns. The weak link in the resolution change, therefore, is the intensifying screen.

in combination with fast-speed screens—the longer exposures result in more photons to produce the image. This increase in x-ray quanta minimizes the mottle effect. A more detailed explanation of all the factors concerning radiographic mottle is left to the physicist.[19]

Since it has been established that intensifying screen speed is generally achieved by the increase of the thickness of the layer of crystals rather than by an increase in the size of the crystals themselves, a fast-speed screen should produce more image unsharpness than a slow-speed screen. In actual practice the use of a fast-speed film with medium-speed intensifying screens is recommended rather than the use of a medium-speed film with a fast-speed intensifying screen. In theory the image transfer effect of the medium-speed screen should be superior to that of the fast-speed screen. The medium screen-fast film combination is preferred to the fast screen-medium film combination. If high contrast films are desired, the use of fast-speed film with fast- or very-fast screens with as low a kilovoltage range as possible is ideal. The lowering of the kilovoltage results in a decrease of scattered radiation, with a decided improvement in contrast.

Utilization of Various Speeds of Intensifying Screens

Different speeds of intensifying screens permit the use of a variety of new x-ray technics. Using intensifying screens or films of several different speeds in an x-ray department can be confusing. Many departments desiring to experiment with different speeds of intensifying screens purchase new cassettes for the installation of the new screens. Even if the additional expense of a large number of new cassettes is not a consideration, the storage and handling of large numbers of new cassettes can present a problem. New cassettes are often purchased because their different appearance makes it easy to distinguish them from existing departmental cassettes, and so alerts the staff to speed differences. The old cassettes contain the original screens, and the newer cassettes house the newer screens.

Different cassette fronts may absorb radiation to different degrees. Variations can be as high as 5 kVp between a thinner bakelite front cassette and a heavier metallic front cassette. One of the early difficulties encountered with the use of fast-speed intensifying screens can be attributed to this difference in cassette fronts, for most older cassettes were made with bakelite fronts and were fitted with medium-speed screens. When fast-speed screens were used in a heavier metal front cassette and compared directly with medium-speed screens in a bakelite front cassette, the technical results were disappointing. Such an evaluation is based on the assumption that the medium-speed screens are fluorescing at the proper light level which they possessed when originally installed. Frequently, a significant decay has occurred in the light output of the original screens due to normal aging, and therefore it is impossible to compare screen speed. If medium- and fast-screens are functioning properly, a reduction of 50 per cent in any given mAs value from the medium-screen technic should result in radiographs of duplicate densities when fast-screens are used. *This is true only if the cassette fronts of both screen combinations are made of the same or similar material.*

If a cassette with a more dense front were used, as much as 5 kVp could be required to overcome the attenuation of the x-ray beam. The substitution of fast-speed screens in bakelite front cassettes when one is accustomed to medium-speed screens in a dense metal front cassette is quite advantageous. Not only can one halve the milliampere-second value, but the added bonus of a reduction of up to 5 kVp can significantly enhance radiographic contrast. Another technical alternative is to retain the original kilovoltage value and reduce appropriately the milliampere-second value.

In order to solve this screen front problem, an easy approach is to use the same make of cassette for all types of screens and to differentiate the speeds of the cassette fronts by color coding. A tinted front cassette is available in a variety of bright colors from an American cassette manufacturer.

For the past 10 years this author has been using a simple and yet inexpensive method of color coding. As existing intensifying screens are replaced, due to aging or damage, the new screens of different speeds are placed in the old but serviceable cassettes. A facing of colored sheet vinyl is glued to the cassette front for the purpose of color coding. Bright colors are necessary, as opposed to the pastels, since cassettes frequently are stored in dark areas—for example, pass boxes, cassette storage carriers, control booths, etc.—and color coding must be rather pronounced to be easily recognized. A conventional black front cassette is used for medium-speed screens. A bright orange vinyl covering identifies the slow-speed screens, which are limited to 8″ × 10″ and 10″ × 12″ for routine extremity studies. A bright yellow vinyl covering is used on the fast-speed screens, available in only the 14″ × 17″ size for radiography of the pregnant abdomen, the obese patient, and bedside technics. Green front cassettes containing fast-speed screens, in violation of the color coding system, are consigned to the operating room. This change in color was made so that the operating room cassettes would not find their way into conventional areas. This vinyl covering system has been in use on the same cassettes for more than 10 years and has outlasted many of the intensifying screens. Some of these cassettes with their original vinyl facings have

had 2 or 3 changes of intensifying screens. The vinyl covering itself, however, tends to fray or damage at its edges.

Multiple Screen and Film Combinations

A simple way of producing 2 radiographs of different density in a single cassette is accomplished by the addition of a second radiographic film. Routine cassettes can be loaded with 2 films of different speeds, one an average-speed film and the other a high-speed film, thus accomplishing a dual-density effect. The average-speed film will be slightly lighter in density than the high-speed film. An increase in technical factors is necessary, depending on the type of film or screens used. Some medium-speed screens have a stained slower front screen and a faster back screen; therefore the faster-speed film should be placed in contact with the posterior screen to secure one film of adequate density (Fig. 1-11).

Multiple screen combinations can be used to demonstrate specific differences in density in any given area of the body. If medium-speed screens of approximately the same gain value are used and an additional medium screen is added to the cassette, we have a package or combination consisting of a pair of medium-speed screens and a single medium screen. By placing a film between the pair of screens and a film between the back of the added screen and the existing mounted screen, a dual-density capability is achieved. By using this screen-film combination, the following technics can be attempted:

Area Under Examination	Dual Screen Film	Single Screen Film
Pregnancy	Fetal skeleton	Placenta (soft tissue)
Chest	Lung and spinal column, behind the heart	Fine peripheral vascular markings
Ribs	Penetration of overlapping ribs	Ribs superimposed on darker lung areas
Spot film of lower lumbar--coccygeal area	L-5 to S-1	Coccyx
Foot	Tarsal area	Toes
Bronchography	Main stem bronchus superimposed on the mediastinum	Peripheral bronchial markings
Dorsal spine	Dorsal spine	Ribs adjacent to the dorsal spine

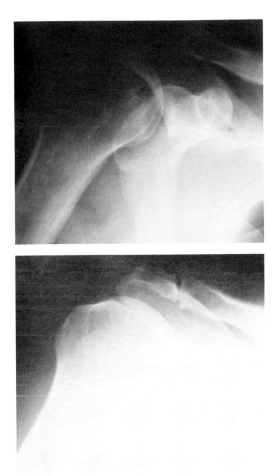

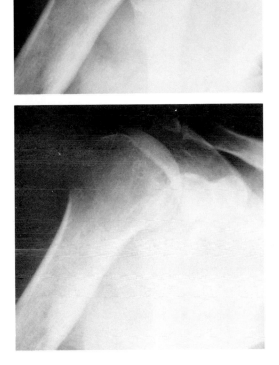

Fig. 1-11. Two films of different speeds in the same cassette. A regular-speed and a fast-speed film was placed in contact with the posterior intensifying screen. The 2 left radiographs were made by using a medium-speed pair of screens with a slower front screen and a faster posterior screen. Note the significant change in film density.

The 2 right radiographs were made with a medium-speed pair of screens of similar speed. The density difference is barely noticeable.

The use of a dual-density technic was reported by radiologists involved in the original cardiac transplant procedures. Two screens of different speeds were used to produce films of a lighter and a darker nature. One of the chest films on the postcardiac-transplant pa-

tient was of a conventional density and the second film of a significantly darker density, a sort of overpenetrated film. The author states that this dual-density technic was quite helpful from a diagnostic standpoint.[17]

Body Section "Book" Cassettes

Screen "book" cassettes for simultaneous multilevel exposures during tomography are available from several manufacturers. Manoel de Abreu of Rio de Janiero developed this radiographic technic known as *simultaneous tomography,* in 1947. In November of 1948 he described his methods to procure several radiographic cuts of reasonably uniform density and detail with a single exposure to the patient.[10] Related technical information will be found in Chapter 8.

Graded or Compensatory Speed Intensifying Screens

Compensatory intensifying screens can equalize differences in patient densities. These units can vary in speed from length to width or from side to side. The speed of the screen is increased by layers of crystals which have been poured in a wedge-like fashion; a slow-speed effect is achieved at one end by the use of a thinner layer of crystals, and a fast-speed effect is created at the other end by a thicker layer of crystals. Such grading or variation of screen speed replaces the use of a compensating filter at the x-ray tube. This type of screen can be used for full-length lower extremity arteriography, since the thicker pelvic area of the patient and the thinner ankle area must both exhibit an equal radiographic density. A major disadvantage of this technic is the exposure of the patient to the full effect of the x-ray beam, for although the thinner areas of the patient appear lighter on the radiograph, this is due to the speed of the screen rather than to the lessening of exposure.

With conventional screens the use of a filter at the source to attenuate the primary beam overcomes this problem. Although a compensatory screen can be easier to use than a compensatory filter with regular screens, it is extremely difficult to justify using this type of screen for taking radiographs of the pregnant abdomen.

Some Common Intensifying Screen Problems

Separating Older and Newer Screens. If one is contemplating the purchase of new intensifying screens, it is necessary, particularly in a large department, to separate the newer screens from those that are older and yet satisfactory. A distinct difference in speeds may be noted between the older and newer screens. The older screens should be transferred to a specific area, such as the fluoroscopic suite, the accident ward, etc., and, to avoid mixing up the older screens

with the newer ones, the cassettes should be clearly marked so that they can be visually identified.

Surface Staining. As some types of intensifying screens age, they tend to stain. This surface stain can vary in color but is generally tan or beige. Such undesirable staining of an intensifying screen produces much the same effect as that of the previously described slow screen. Intensifying screens of any speed can deteriorate excessively when staining occurs. In some instances a 20 kVp drop in screen speed has been noted by this author. Representative radiographs demonstrate this loss of screen speed (Fig. 1-12). Surface staining reduces the light given off by the fluorescing crystals to a point of no return. *Screens which are identical in nature and were installed on a given date can still vary in screen speed over a period of time.*

Evaluation. During routine inspection of intensifying screens the inspector may occasionally become "snow-blind." The repetitious evaluation of hundreds of large white reflective surfaces tends to pale vision and therefore judgment. In a recent visual inspection of approximately 500 cassettes, all were termed "adequate" for routine departmental use. When these same cassettes were evaluated with x-radiation, 75 were found to vary significantly in light output. The oldest screen in the lot was not actually the slowest screen in the grouping. Surface staining had occurred in some of the newer screens, with resulting screen gain loss. *It is therefore important to stress that screens be evaluated by the use of radiation.*

Control films should be made from a half-dozen screens that are of new or recent installation. Radiographic film of the same film emulsion number must be used for the entire evaluation. Exposures should be of a reasonable length ($\frac{1}{4}$ of a second or longer) to avoid variation in timer output, and a fixed collimator opening should be used for all cassette sizes. Lead diaphragming at the table level should be used to avoid undercutting of the radiographic phantom or step wedge by scattered radiation. It must be assumed that

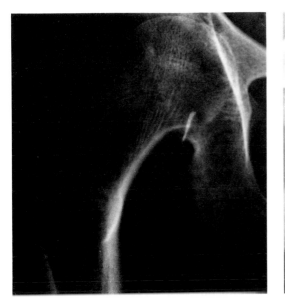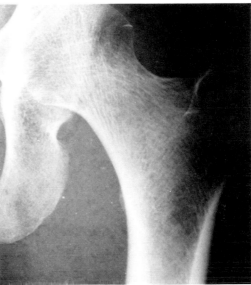

Fig. 1-12. The effect of surface staining on the speed of an intensifying screen. Two 8″ x 10″ cassettes with screens of the same speed were placed side by side in a Bucky drawer. A 14″ x 17″ projection of a pelvic phantom was made in the AP position. Note the difference in film density between the right hip and the left hip. Surface staining of the intensifying screens on the right has diminished the speed of the screens. Screen testing should include the use of x-radiation to guarantee screen speed, for some of the surface change can be difficult to detect by visual observation.

a properly calibrated radiographic unit is used, and that controlled processing conditions are available. In any average large group of screens, even when the finest control is available, some of the screens seem to vary slightly in speed. These variations in film density can be due to many things, such as difficulties with processing, timing, or basic output variations in the x-ray machine.

Intensifying Screen Artifacts

The major purpose of screen cleaning is to eliminate surface marks which result in film artifacts (Fig. 1-13). In many cases the collection of surface dust found on intensifying screens precludes their use for the localization of small, opaque foreign bodies. Dirt or dust artifacts are particularly annoying in evaluation of the eye for evidence of a metallic foreign body. An interesting technic

to avoid this problem is to utilize either nonscreen film or screen film in a cardboard holder. High kilovoltage can be used in conjunction with a Bucky to shorten exposure time. The heavy contrast radiograph usually associated with Sweet's localization method is avoided when nonscreen film is used.

Cleaning Intensifying Screens

It is common practice to spray intensifying screen cleaner directly on intensifying screens rather than into a cleaning towel or tissue. Since many screens are quite porous, this can be a bad practice. One should spray a minimum amount of cleaner directly on the cleaning cloth, and then apply this slightly dampened cloth to the screen area. Excessive spraying can result in the absorption of the cleaner by the felt lining of the cassette, with possible subsequent damage to the screen.

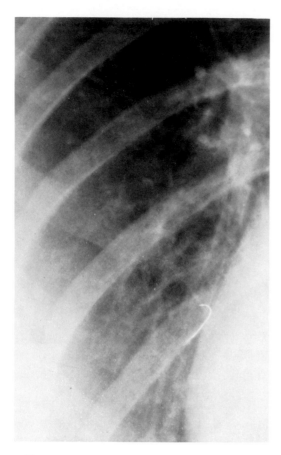

Fig. 1-13. Unusual artifact. Metallic-like object, resembling a fishhook (barbed end of hook, and a needle-like eye).

This opaque-like object does not represent a foreign body on or within the chest walls but rather an artifact on the surface of one of the intensifying screens used to make this exposure. A small thin piece of black thread was found in this cassette, lying in a hook-like position. The thread was frayed at one end, giving the illusion of a needle-like opening on the finished radiograph. This black thread prevented the light from one of the fluorescing screens from reaching the film, producing a metallic-like artifact.

FLUOROSCOPIC SCREENS

The basic principle of image detection via a fluoroscopic screen was popularized by Thomas A. Edison, who actually invented and coined the term *"fluoroscope"* in March of 1896.[4-6]

When fluoroscopic screens are activated by radiation, they glow yellow-green within the range of the wavelength to which the dark-adapted eye is sensitive. Because of this luminescent effect, the visible rays can be directly detected by the eye.[4]

The standard fluoroscopic screen has a 2- to 3-line per millimeter resolution capability as compared to the 12-line per millimeter resolution capability of a currently available medium-speed intensifying screen. Despite an increase in crystal size, with a resultant gain in visible light, the fluoroscopic image is still quite dim. When the brightness of the fluoroscopic screen is compared with the brightness level at which a typical radiographic film is interpreted, the gain factor of the film is about 10,000 times that of the fluoroscopic screen. This means that a typical fluoroscopic image is 1/10,000 of the brightness level of the typical viewed radiograph.[15] Therefore it is important that the eye be dark-adapted as a prerequisite for successful viewing of the dimly illuminated fluoroscopic image.

One of the earliest reports of the need for dark adaption of the eye prior to fluoroscopy was that made by Bergonié and Carrière. Many early pioneers made similar observations and a variety of colors were suggested for use in dark-adaptation lenses. The use of the red-adaptation goggle was first proposed by Trendelenburg in 1916.[29] Fluoroscopy even after adequate dark adaptation of the eye is more time-consuming than fluoroscopy by means of an image-intensification device, which will be discussed in Chapter 10. Without proper dark adaptation one will unnecessarily and unjustifiably overexpose both patient and operator to the higher intensities of radiation required to produce the brighter screen image for visualization.

The basic crystals utilized in the manufacture of fluoroscopic screens are zinc cadmium sulfide. These crystals, which are approximately 30 microns in size, are substantially free from afterglow or phosphorescent effect. Like the crystals of intensifying screens, these zinc cadmium sulfide crystals absorb energy in one form and emit this energy as visible light. Fluoroscopic crystals are quite large as compared with standard intensifying screen crystals, the ratio being about 6 to 1 (30 microns to 5 microns) in size.

Application of Fluoroscopy

The fluoroscope is used to observe the dynamics of internal organs. It is much more economical to operate than conventional radiographic filming. Although fluoroscopy is outstanding for motion or detail search, radiographs are necessary for good detail resolution and afford greater viewing time. The fluoroscope is particularly useful in patient positioning and as a centering aid in conjunction with the spot film device. With the fluoroscope the object under study can be observed for a considerable length of time and in a variety of positions. A definite 3-dimensional effect can be achieved by rotation of the patient during fluoroscopy. Rotational devices (cradles) are available for rotational positioning technic.

Since radiation must be used for the entire viewing time during fluoroscopy, the amount of radiation to which both patient and operator are exposed is a matter for concern unless an image intensifier can be used in combination with a television and videotape recorder. The use of the videotape recorder or videodisc scanner gives to the fluoroscopist the advantage of an instant replay.

Greater viewing time, however, is made available to the physician by the use of properly exposed radiographs, which exhibit excellent detail as well as fine resolution.

Fluoroscopic Spot Film Devices

The first commercially produced and instantly successful spot film device was made by Frank Scholz of Boston about 1930.[11] For many radiologic technologists who trained or practiced during the '30's or '40's the name of the Scholz tunnel was synonymous with the fluoroscopic spot film device, for almost every available piece of heavy-duty fluoroscopic equipment came equipped or was adaptable to the Scholz spot filmer. The spot film device produces a "still" radiograph of a moving anatomic part. The radiologist can

at any time during the fluoroscopic examination take a single spot film or multiple views, as many as 4 per film, using precise positioning technics. The fluoroscopist actually sees what he is about to radiograph. Future use of a spot film device coupled with an image-intensifier-television unit includes a positioning device to be utilized by the radiologic technologist. For example, spot films can be exposed of the gallbladder or kidney area after first localizing these areas with an intensifier.

There is a growing trend to utilize specially trained radiologic technologists to relieve the radiologist of his fluoroscopic duties. Routine fluoroscopy can be performed by the radiologic technologist with an image-intensifier television system. The televised fluoroscopic image can be linked to a tape recording device for the convenience of the radiologist who would like to view the fluoroscopy at a later time. Conventional radiographic spot films can be taken during the examination by the technologist, and kinescopic or videodisc scanner "stills" can be made by the radiologist. Using a spot film device as a positioning aid for direct roentgenographic enlargement technic was suggested by Isard, Ostrum, and Cullinan in the early 1960's. This technic requires a spot film device and an image enlargement radiographic tube known as a "fractional-focus" x-ray tube.[14] Representative radiographs are shown in Chapter 4 to demonstrate this positioning technic.

Fluoroscopic-Radiographic Dose Comparison

It is quite easy in using a fluoroscope to expose a patient inadvertently to radiation for a long interval of time, for the live visualization of an organ in motion can be both fascinating and interesting. The skin of a patient may typically receive about 5 rads during each minute of fluoroscopy in contrast to the dosage delivered during a typical chest radiograph, which is about $1/25$ of a rad. A patient receives more dosage to his skin during a single second of fluoroscopy than he receives from a chest radiograph.[25] A typical distance for a fluoroscopic tube is about 18 inches from the tabletop as compared to a conventional focus-film distance for radiographic films of 40 inches, with chest radiographs having a 72 inch focus-film distance (Fig. 1-1). *A fluoroscope should not be used for an examination that can be accomplished with radiographic film.*

Although fluoroscopic tubes are closer to their image detector than the tube cassette system and use shorter exposures per film, they are still easily damaged by prolonged fluoroscopy or the repetitive making of spot film radiographs. Many fluoroscopic spot film devices make use of phototiming systems, and often tube rating charts are ignored by the fluoroscopist.

Photofluorographic Unit

A photofluorographic unit is a quick, efficient way to accomplish mass survey chest radiography. Photofluorography has been invaluable in the detection of diseases of the chest. Its role as a detector of primary tuberculosis more than justifies its place in the history of radiographic equipment. A photo-

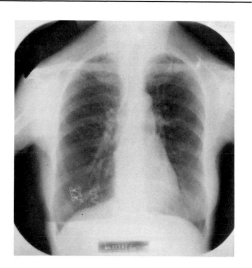

Fig. 1-14 Photofluorogram. Actual size, 70 mm. This photofluorographic study of a female adult in the PA position was taken from a full-size fluoroscopic screen by a camera using single emulsion photographic film.

fluorograph is a survey examination, not a complete roentgen examination, and is never considered as such.

Ever since the discovery of x-ray an attempt has been made to utilize the roentgen image in conjunction with photography to develop an economical and efficient method for mass radiography. Although this process became workable in the '30's, the use of the photofluorographic unit was firmly established by the military during the Second World War. The major technical achievement that popularized the photofluorographic unit was the use of an automatic timer to control film density. This phototimer, developed by Morgan in 1943, made possible an extremely high quality of photofluorographic film.[16]

A photofluorographic unit consists of a single fluoroscopic screen in a lightproof holder. A camera is used to make a photographic image of the full-size fluoroscopic screen (Fig. 1-1). The camera records the details from the screen on photographic film of smaller size, the most popular sizes being 70 mm strip film or a 4″ × 4″ cut film. Because of this photographic reduction of the fluoroscopic image from a 14″ × 17″ area to a 70 mm film, a magnifying lens or a slide projector device is generally used to view the finished film. The fluoroscopic screen of a photofluorographic unit can be composed of zinc sulfide crystals, which fluoresce blue when activated by x-ray, or of zinc cadmium sulfide crystals, which fluoresce yellow-green when activated. Appropriate blue-sensitive or yellow-green-sensitive, single-coated photographic films are matched to the proper fluoroscopic screen.

The basic differences between photofluorography and dual intensifying screen technics are significant. A photofluorographic unit utilizes a single fluoroscopic screen combined with a single emulsion photographic film to take a miniature photograph (Fig. 1-14). Dual intensifying screens are used with a single sheet of medical x-ray film coated on both sides with a photographic emulsion. The dual screen radiograph technic is either actual size or larger than the area being evaluated, whereas the photographic image from the photofluorographic unit is significantly reduced in size.

The photofluorographic unit described above is the most simple system available. Newer mirror optical systems offer improved image definition as well as increased fluoroscopic gain.[26] Even under the best circumstances a photofluorographic unit requires at least 3 to 5 times the exposure required with a double emulsion film and intensifying screens.[18]

REFERENCES

1. Bruwer, A. J.: Classic Descriptions in Diagnostic Roentgenology. vol. 1, p. 98. Springfield, Ill., Charles C Thomas, 1964.
2. Cahoon, J. B.: Formulating X-Ray Tecniques. ed. 7. College Station, Durham, N. C., Duke University Press, 1970.
3. *Ibid.,* p. 117.
4. Chamberlain, W. E.: Fluoroscope and fluoroscopy. Radiology, *38*:383–412, 1942.
5. Dewing, S. B.: Modern Radiology in Historical Perspective. pp. 74–75. Springfield, Ill., Charles C Thomas, 1962.
6. Edison, T. A.: The Edison Fluoroscope. The Electrical World, *27*:360, 1896.
7. Fuchs, A. W.: Principles of Radiographic Exposure and Processing. ed. 2. Springfield, Ill., Charles C Thomas, 1958.
8. *Ibid.,* p. 161.
9. Goodwin, P. N., Quimby, E. H., and Morgan, R. H.: Physical Foundations of Radiology. ed. 4, p. 107. New York, Harper and Row, 1970.
10. Grigg, E. R. N.: The Trail of the Invisible Light. p. 444. Springfield, Ill., Charles C Thomas, 1965.
11. *Ibid.,* p. 414.
12. Handee, W. R.: Medical Radiation Physics. p. 451. Chicago, Year Book Publishers, 1970.
13. Horner, R. W.: Different ways of checking poor screen contact. Dupont X-Rays News, No. 57. Wilmington, Del., E. I. du Pont de Nemours, 1962.
14. Isard, H. J., Ostrum, B. J., and Cullinan, J. E.: Magnification roentgenography. Med. Radiogr. Photogr., *38*(3):92, 1962.
15. Meredith, W. J., and Massey, J. B.: Fundamental Physics of Radiology. p. 341. Baltimore, Williams & Wilkins, 1968.
16. Morgan, R. H.: The automatic control of exposure in photofluorography. Public Health Rep., *58*:1533–1541, 1943.
17. Palmer, P. E. S.: The radiologist and heart transplants. Electromedica, *4*:122–130, 1969.
18. Report of the Medical X-Ray Advisory Committee on Public Health Considerations in Medical Diagnostic Radiology (X-Rays). p. 10. U.S. Department of Health, Education, and Welfare; Washington, D.C., U.S. Government Printing Office, October 1967.
19. Seeman, H. E.: Physical and Photographic Principles of Medical Radiography. pp. 89–104. New York, John Wiley & Sons, 1968.
20. Selman, J.: The Fundamentals of X-Ray and Radium Physics, ed. 4, p. 242. Springfield, Ill., Charles C Thomas, 1970.
21. Sensitometric Properties of X-Ray Films. p. 11. Rochester, N.Y. Eastman Kodak Company, [1963].
22. Some Physical Factors Affecting Radiographic Image Quality: Their Theoretical Basis and Measurement. p. 10. U.S. Department of Health, Education, and Welfare, No. 999 RH 38. Washington, D.C., U.S. Government Printing Office, 1969.
23. Stanton, L.: Basic Medical Radiation Physics. p. 261. New York, Meredith Corporation, Appleton, 1969.
24. *Ibid.,* p. 78.
25. *Ibid.,* p. 74.
26. Ter-Pogossian, M. M.: The Physical Aspects of Diagnostic Radiology. pp. 341–381, New York, Harper and Row, 1967.
27. The Care and Use of Cronex XTRA LIFE Intensifying Screens. p. 9. Wilmington, Del., E. I. du Pont de Nemours, 1966.
28. The Fundamentals of Radiography. Rochester, N.Y., Eastman Kodak Company, 1960.
29. Tuddenham, W.: Dark adaptation. *In* Bruwer, A. J. (ed.): Classic Descriptions in Diagnostic Roentgenology. vol. 1, pp. 741–747. Springfield, Ill., Charles C Thomas, 1964.
30. Van der Plaats, G. J.: Medical X-ray Technique. ed. 3, p. 194. Eindhoven, Netherlands, Philips Technical Library, 1969.

2. Control of Secondary Radiation

There are more variables than constants in the practice of radiologic technology. Patient size, positioning technics, and tube angles often vary but the grid ratio remains the same in a given situation, while field size, whether determined by a cone or a collimator, is frequently ignored. The value of an acceptable ratio grid in combination with *primary beam* limitation will be stressed in this chapter. This material is intended to interest the radiologic technologist in the many ways to reduce or control *secondary radiation*. It is not important to stress the nature of *secondary radiation;* a detailed explanation is left to the physicist.

SOME TECHNICS AND DEVICES

Scattered Radiation

Although scatter radiation affects the overall film, we find that the density of the darker areas of the radiograph is such that the additional density is not detrimental. In areas of average or low densities, such as opacified blood vessels, anatomic details can be lost or obscured by the overlying supplemental density. Fog does not damage an entire film equally. It is particularly damaging to the high contrast or white areas of a radiograph.

All supplemental densities cannot be blamed on scattered radiation. Many conditions produce a fog-like image on the finished radiograph. These include careless handling of film under improper safe lights, film processing difficulties, storage conditions, improper collimation, a low ratio grid combined with high kVp/low mAs technics, etc. Any of the above situations can add a veil of fog to the low density areas of a radiograph. Scattered radiation is probably the biggest single factor contributing to poor diagnostic quality, for it reduces the contrast between visible areas of detail on the radiograph. Early pioneers were dependent on long extension-type cones to limit effectively the primary ray and therefore reduce the amount of scattered radiation. Although it is true that pioneer technologists used slower emulsions which were less sensitive to primary or secondary radiation, they did effectively reduce the amount of scatter radiation generated by proportionately reducing the area being exposed. Many technologists feel that we expose these small fields with radiation dosage in mind. While this is true, it is important to stress that if x-radiation were completely harmless, and if film size and the amount of radiation given to patients were of no importance, it would still be necessary to restrict the primary beam size to improve the quality of the radiographic image. *The most important way to reduce scattered radiation is to limit the area being exposed to primary radiation.* It is senseless to depend completely on the clean-up ability of the grid or Bucky for the purpose of eliminating scattered radiation. The value of collimating a primary beam cannot be overstressed.

Since higher kilovoltage values generate more scattered radiation, it follows that moderate kilovoltage ranges should be utilized with moderate ratio grids. It is important not to lower kilovoltage values significantly, for this lowering of kVp, with the resultant compensatory increase in milliampere seconds, creates a new problem—an increase in the generation of heat units (see Chapter 3). *The lowering of kilovoltage and the raising of milliampere seconds to lessen scattered radiation is a dangerous substitute for proper collimation combined with a proper ratio grid.*

Another form of supplementary film density is *back-scatter.* Backscatter arises from the front or back of the cassette or the table itself. The lead foil backing of the cassette helps to attenuate backscatter.

The generation of an x-ray beam is a complicated matter, and to overstress material of this nature is not the purpose of the author in this textbook.

Primary radiation is the radiation that is emitted from the x-ray tube. It enters the patient at a predetermined field size, depending on the method of collimation, passes through the area being examined, and is absorbed or attenuated in varying degrees. This attenuation varies with the thickness and density of the part under study. For example, bone absorbs significantly more radiation than does flesh. Unabsorbed, *remnant radiation* exits from the object and is received by the image detector. Scatter radiation is generated, depending on the composition of the part being examined and the kilovoltage level being employed. The amount of scatter radiation increases as one increases the size of the area to be examined—for instance, from an 8″ × 10″ to a 14″ × 17″ field size. Scatter radiation generated by an increase in field size and kilovoltage is almost impossible to overcome with a low-ratio grid. In the area of 15 to 20 cm of a body part with a full field port, the intensity of scattered radiation at the film is 2 to 4 times greater than the intensity of the primary beam. If scattered radiation is reduced to approximately 20 per cent or less of the total beam intensity that reaches the detection device, a radiograph of good quality should result.[15]

Air-Gap Technics

Scattered radiation is extremely diffuse and is emitted in all directions. An extremely descriptive word for scattered radiation would be "spacial."[5] Because scattered radiation diffuses in all directions, it is generally more oblique in nature than primary radiation and must travel through longer paths of the body. These longer routes help to weaken or attenuate scattered radiation. Since the patient is the major source of scattered radiation, it would be ideal if the film could be moved some distance from the patient to take advantage of this "spacial" effect. In Chapter 4 an air gap is used to advantage with direct roentgenographic enlargement technic.

Many departments use a non-Bucky or non-grid technic for radiography of the lateral cervical spine. Why is it possible to radiograph the cervical spine in the lateral position without a grid or Bucky but necessary to use a grid or Bucky for the AP film? The "spacial" effect of scattered radiation is the answer, for with the AP projection, the x-ray film is almost in complete contact with the part, whereas in the lateral view the object-film distance is from 8 to 10 inches, resulting in an air gap. Scattered radiation is disseminated in all directions, with little or none reaching the image detector. Scattered radiation that is parallel to the central ray is somewhat weakened by the increase in object-film distance (Fig. 2-1).

An interesting use of the air-gap technic is the high kilovoltage non-grid examination of the chest. Moving the patient 6 inches from the cassette will cause the intensity of the scattered radiation to vary inversely with the square of the new and increased object-film distance. A focal object distance of 10 feet or more must be used to overcome any magnification of the contents of the chest. Kilovoltage values as high as 150 kVp are used with extremely short exposure times. A well collimated primary beam is mandatory.

Primary Ray Restricting Devices

A collimator or a good-quality cone is one of the most important radiographic accessories. Its value must be weighed seriously against the value of a grid or Bucky, for since the grid is rarely changed, one learns to live

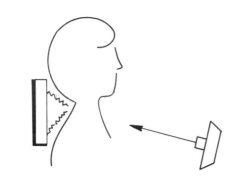

A   MINIMAL O F D — 40" F F D
      A P CERVICAL SPINE

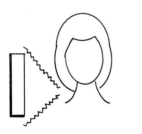

B   AIR GAP
      INCREASED O F D — 72" F F D
      LAT. CERVICAL SPINE

Fig. 2-1. "Air-gap" effect. It is possible to use an air-gap to reduce *scattered* radiation when the cervical spine is examined in the lateral position because of the increased object-film distance, as compared with the AP projection. Scattered radiation is disseminated in all directions and is weakened by the increase in object-film distance.

with its scattered radiation clean-up capability. A collimator helps to make the grid more efficient, for the thicker the part being radiographed the greater is the need for a reduction in portal size. Since there is little or no scattered radiation present when an object as small as a wrist is radiographed, a grid is of no value. The use of a grid for a small object would not be considered by most technologists, and yet many technologists use a high-ratio grid for extremely thin objects. There is a distinct disadvantage in using a high-ratio grid with a thinner body part: the patient receives more exposure to radiation

than is necessary for a diagnostic examination.

Grids are essential in examining heavier body parts. A general rule to follow is that any anatomic part 10 cm or thicker of relatively dense tissue requires the use of a grid or Bucky. The major exception is the adult chest. Here, due to the radiolucent nature of the lungs, significantly less scattered radiation is produced than in the more dense abdomen.

Various devices are used to restrict the primary ray, including primary source diaphragms, conventional and telescopic cones, external lead diaphragms, placed on or near the patient, and the newer light beam collimators. A cone or collimator restricts the area under study to a fixed field size. The use of a cone that will limit primary radiation to a smaller field size than the film in use results in a "cone" cut. The visualization of a cone cut on a radiograph is not an error but rather a technical compliment, for it proves that efficient limiting of the beam was accomplished.[2] Not only is the value of a cone far greater than most technologists realize,[2] but greater than they care to accept, for poor primary ray limitation is one of the most serious technical problems encountered.

Primary beam collimation by means of a single diaphragm is seldom used. The use of a single diaphragm, although not as efficient as the use of a cone or collimator, can be fairly effective if used in close proximity to the x-ray source (Fig. 2-2).

Radiographic Collimators

The first radiographic collimator, produced by the Howdon Videx Corporation of Mount Vernon, New York, was termed an "adjustable cone."[8] Collimators simultaneously adjust several pairs of lead shutters so that they act synchronously with a beam-defining light to restrict the primary beam. Most collimators have common features. These include a very high-intensity light source, coupled with a mirror reflection system, which is so arranged that the light field indicated on the patient matches the x-ray field size as closely

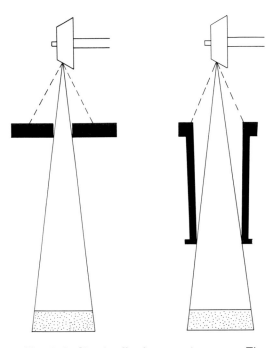

Fig. 2-2. Single diaphragm at source. The use of a single diaphragm at source can be as effective as an extension cylinder with a poor primary diaphragm. The ideal cone uses both principles, collimation as close to the source as possible and as close to the patient as possible.

be controlled by a poor-quality collimation system.[12]

Off-focus or *stem radiation* has a value of 8 to 10 per cent of the primary beam.[13] Ter-Pogossian suggests that off-focus radiation may represent up to 25 per cent of the on-focus radiation.[18] The best answer to the control of off-focus radiation, as well as collimation of the primary beam, is a series of lead shutters at several levels, with the initial shutters in close contact with the window of the x-ray tube and the final series of shutters as close to the object as possible.

Some light beam collimators have the first set of shutters exiting from the collimator into the tube port, lessening off-focus radiation. If one has a collimator of this type and is still having difficulties with off-focus radiation, it is important to check whether the uppermost diaphragm is in its proper position. It is quite simple to remove these diaphragms at the time of installation if one is experiencing difficulties in alignment of the primary ray and light beam. Unfortunately,

as possible (Fig. 2-3). An external "cross-hair" marking indicates the position of the central ray, taking the guesswork out of centering and localizing.

Light beam collimators consist of several sets of lead diaphragms mounted at different levels. One of the sets of diaphragms is mounted preferably as close to the tube port as possible. This closeness of the uppermost diaphragm sharply collimates the primary beam at its source, eliminating unwanted *off-focus* or *stem radiation*. Often when the technologist is using a cone or light beam collimator, he sees radiographic images extending beyond the predetermined port size (Fig. 2-4). This effect can be caused by the radiation produced when electrons strike metal surfaces other than the target area of the tube. This unfocused radiation exits at a variety of angles from the tube and cannot

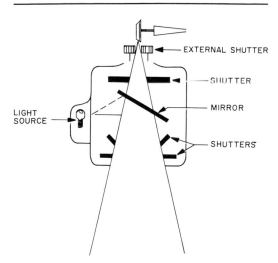

Fig. 2-3. Modern light beam collimator. Several sets of lead diaphragms are mounted within the collimator housing at different levels. The first set of shutters should be mounted as close to the tube port as possible to help to eliminate *off-focus* or *stem* radiation. A light source is used with a mirror to indicate the x-ray field size as closely as possible.

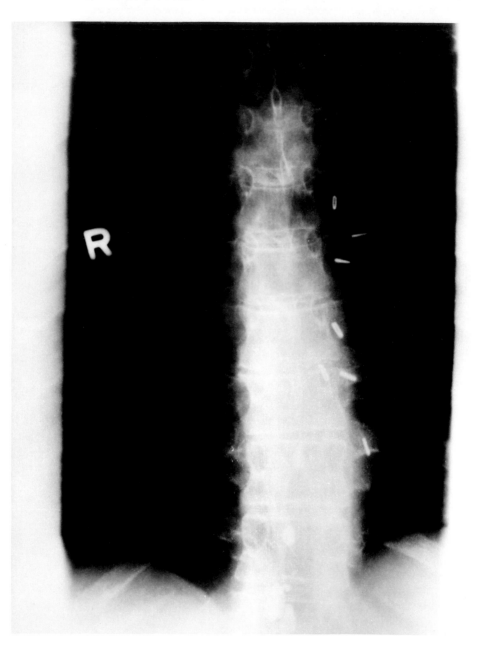

Fig. 2-4. Off-focus radiation. Note radiographic details beyond the collimated image. These details could be the result of poor collimation of the x-ray beam at its source. Off-focus or stem radiation can exit at a variety of angles from the x-ray tube port, and it is important that the first series of lead shutters of the collimator be as close to the tube window as is possible to minimize the effect of off-focus radiation.

it is easier to remove the primary diaphragms than it is to spend time trying to make them function correctly. *The removal of the first set of shutters makes the collimator virtually useless.* All of the lead shutters of a collimator must be in perfect alignment. The premature closing of an upper set of lead shutters negates the collimating ability of the remaining lead shutters, producing an unsharp penumbral effect on the finished radiograph (Fig. 2-5).

Radiographic Cones

Prior to the introduction of the collimator most radiologic departments worked with 3 or 4 sizes of radiographic cones. These cones had a fixed field size at a given distance, either a 36 or a 40 inch focus-film distance. The cones were referred to as a 14 × 17 inch cone, a 10 × 12 inch cone, an 8 × 10 inch cone, and an "extension" cone. The long extension cyclinder varied from approximately 10 to 20 inches in length for spot radiography (Fig. 2-6). The designation "14 × 17 inch cone" meant that this cone at the proper distance would cover a 14 × 17 inch radiographic film without a significant cone cut.

Prior to the 1960's it was relatively easy to make an adjustment in technic for a change in cone size. When a smaller cone, such as an 8 × 10 inch size, was substituted for a larger cone, an approximate increase of 5 kVp over that of the previously satisfactorily exposed radiograph would be necessary. For example, a radiograph of the abdomen requiring 50 mAs at 70 kVp on a 14 × 17 inch film would require 75 kVp at the same mAs value for an 8 × 10 inch radiograph if the production of a similar radiographic density was desired. The use of the extension cylinder at maximum length for a spot film of the right upper quadrant (of the same abdomen) required an approximate 5 kVp increase over the adjusted technical factors of the 8 × 10 inch film, or a 10 kVp increase over the original 14 × 17 inch study. Density variations due to cone size were easy to overcome with only 4 field sizes to choose from.

Consider the plight of the present-day

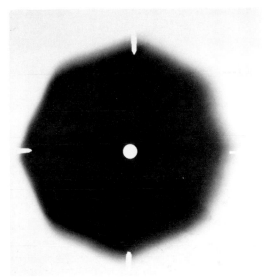

Fig. 2-5. Primary diaphragm collimation. The use of a primary diaphragm or primary lead shutters as close as possible to the x-ray tube to avoid *off-focus* radiation is recommended. In multileaf collimators it is important that the shutter levels be in proper alignment to secure maximum *primary* beam restriction.

This radiograph was made with a multileaf collimator prior to proper diaphragm alignment. The primary shutters are closing prematurely, and the secondary shutters are not being used for restriction of the *primary ray*. The unsharp penumbral effect of this circle-like exposure is the result.

---

technologist, who has dozens of possible field size variations with the collimator. With conventional radiographic technics, beam variations can result in serious density differences (Figs. 2-7 and 2-9). Although density changes can occur when the collimator port size is varied during phototiming technics, the phototimer usually will compensate for overcollimation or undercollimation.

Comparison of Technical Features of Cones and Collimators

When new multileaf collimators having variable rectangular shutters are first installed, many radiologic departments are dissatisfied with them. Unfair comparisons

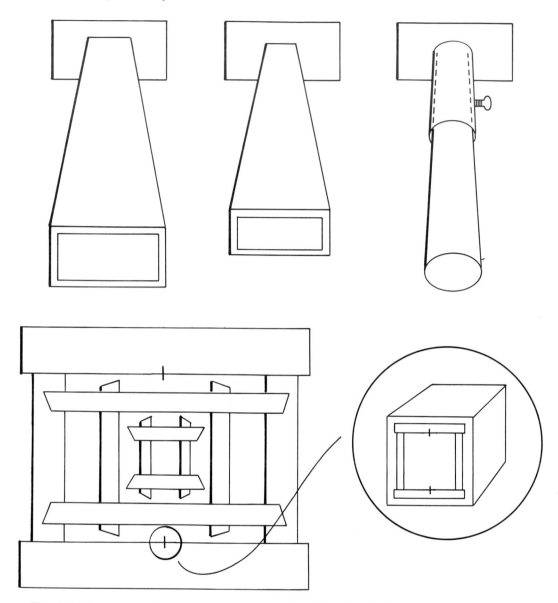

Fig. 2-6. Conventional cones versus multileaf collimator. Typical conventional cones are shown and compared with the variable rectangular shutters of a multileaf collimator. Note the use of a lead marker on the final set of shutters of the collimator to be used as a field size indicator. See Figure 6-4.

are made between the technical factors of the old conventional cones and the new technical factors required with the new multileaf collimator. A new light beam collimator can have an increased amount of inherent filtration. Even if filtration, whether inherent or added, remains the same, there is still a definite difference between the new and old technical factors. This is easily explained, for the port of entry of the primary beam has been significantly reduced by the use of a collimator; therefore, scattered radiation has been more effectively controlled (Fig. 2-8). Along with reduction in radiographic density

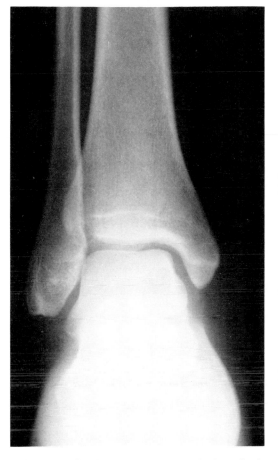 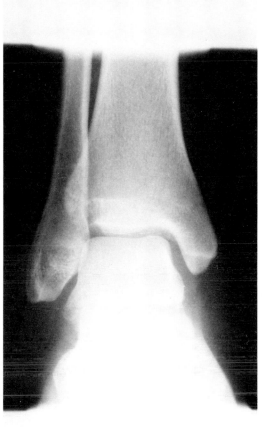

Fig. 2-7. Collimator shutter variation. Serious density differences can result from over-collimation or undercollimation of the primary beam. If relatively little scattered radiation is generated by a thin body part, such as the ankle, density changes are minor. The film of the ankle on the left was made with a 7 x 17 inch beam, whereas the radiograph on the right was sharply collimated. Identical factors as well as the same cassette were used for both exposures. Increased contrast is observed in the well collimated study.

During examination of the skull or trunk, density changes of 100 per cent or more can occur with open shutter technics. Conversely, underexposed films will be the result of an overly restrictive collimating habit.

---

is a definite improvement in radiographic contrast. An increase in technical factors, therefore, must be made if a collimator is to function properly.

Despite the high degree of efficiency of a well-controlled beam size, a collimator is of no avail if the shutters are left open beyond the required field sizes. Too frequently a chest examination is exposed at a 72 inch focus-film distance, with the collimator shutters open to maximum. The field size, if measured, would be over 3 ft × 3 ft square, for a total of 9 square ft of patient or area exposure. If the technologist who practices his profession in this manner is not concerned about exposure to the patient, he should be concerned about image quality, which suffers significantly from this practice.

Telescopic cylinders are available which permit their extension or contraction to limit

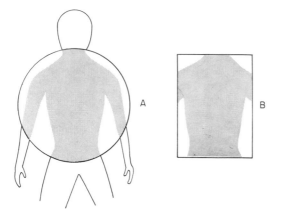

Fig. 2-8. Field size comparison. Old style circular port technics (A) subjected the patient to considerably greater radiation exposure. A rectangular cone or collimator technic (B) helps to overcome this problem. Since the *primary* beam is reduced in size with the rectangular port method, it follows that *scattered* radiation is also decreased, and that there is an improvement in film quality.

the x-ray beam. Although these cylinders have been all but forgotten with the advent of the collimator, many new collimators have external cone tracks which permit the use of a telescopic cylinder. A common error with the utilization of a telescopic cylinder in conjunction with a collimator is the failure to reduce the collimator shutters to the small field size required by the telescopic extension (Fig. 2-9).

Automatic Shuttering Devices

Available are automatic beam limiting devices utilizing sensors in the Bucky drawer to measure the size of the cassette in use.[20] As the Bucky drawer locks are adjusted, sensors automatically measure the cassette size and restrict the primary beam appropriately. If the radiographic tube is raised or lowered, sensors correspondingly restrict or open the shutters, so that for any focus-film distance the exit beam size is exactly the size of the cassette in use. At the end of an exposure, the cassette is removed from the drawer and the collimator shutters automati-

cally reduce to a smaller port size. If special field sizes are required, an automatic override can be used. When the x-ray tube is turned at a 72 inch focus-film distance and positioned so that the x-ray beam is perpendicular to the chest cassette changer, the automatic sensing device restricts the collimator shutters to the proper 14 × 17 inch beam size for teleoroentgenography of the chest.

Automatic Collimator versus the Manual Collimator

The use of an automatic cassette sensing device in conjunction with a multileaf collimator eliminates multifold variations in port sizes. This device combines the ultimate in primary ray restriction with the simplicity of the 3 or 4 cone technics of yesterday. Some of the problems of the conventional collimator eliminated by the automatic collimator will now be examined.

There are technologists who overcollimate or overrestrict the beam size, and technologists who leave the shutters for every examination at a larger size than that required. A common error can be the repeating of a radiograph by a technologist other than the technologist who made the initial examination. Since the 2nd technologist is not familiar with the collimating habits of the 1st technologist, serious density variations can occur. For example, suppose the 1st technologist exposes a 10 × 12 inch film of the lumbar spine and neglects to collimate the beam, electing to leave the port size a full 17 × 17 inches. Let us further assume that the original radiograph is slightly underexposed, and a decision is made to repeat the examination with an increase of 10 kVp to bring the radiograph to proper density. If the technologist who repeats the film is unaware of the previous collimating error and makes the correct technical adjustment of 10 kVp, but restricts the primary ray to the proper 10 × 12 inch field size (120 square inches) versus the original 17 × 17 inch beam (289 square inches), there will be a significant decrease in the amount of scattered radiation. The 2nd radiograph, while exhibiting more

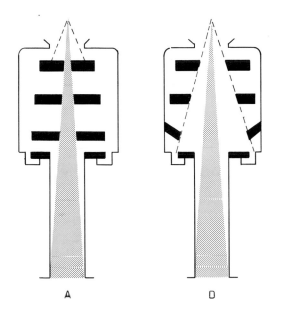

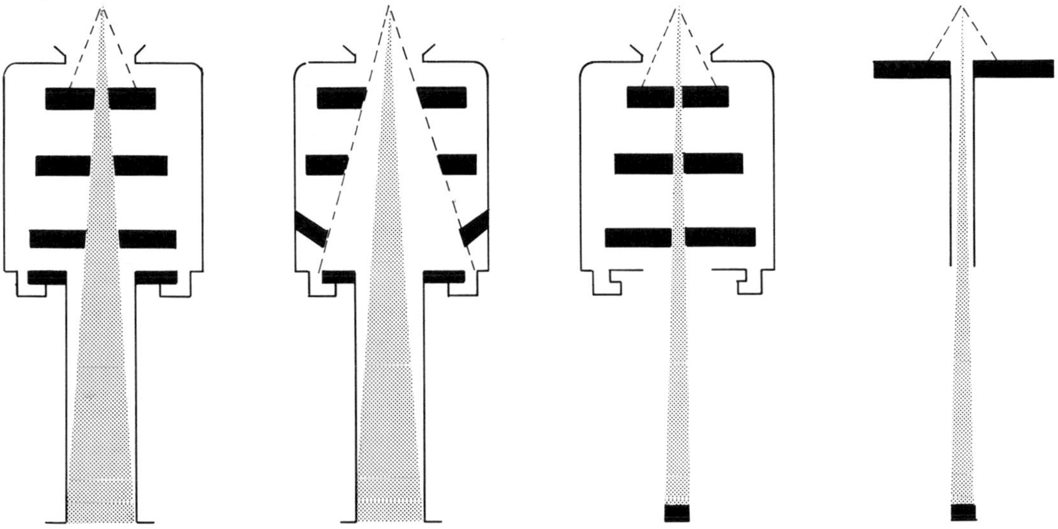

A   D   C   D

Fig. 2-9. Collimator shutter. Use and abuse. Many modern collimating devices with multiple lead shutters come equipped with an external track for the use of an extension type cylinder cone.

Figure A shows the proper use of the extension cone by the restriction of the shuttering device to the size of the extension cone for maximum efficiency.

Figure B shows a typical error made in utilizing an extension cone or a final lead diaphragm in the external track. Since diaphragms and extension cones are opaque to light as well as x-ray, it is impossible to know whether the internal lead shutters are restricted to the required size, as illustrated in Figure A.

Supposedly high-detail radiographs of areas such as the gallbladder are made with the extension cone at maximum length, and the radiographs do not exhibit technical improvement over conventional films. The solution is simple: collimate the shutters as in Figure A, for in Figure A with proper shutter restrictions the 3 lead shutters of the light beam collimator as well as the final shutter effect of the extension cone are used for maximum beam restriction.

In Figure B use is made only of the final shuttering effect of the extension cone for single shutter collimation of the primary beam. This violates every principle of collimation.

Figure C. There is an unnecessary tendency to restrict the shutters of a light beam collimator to extremely small port sizes in examining fingers, toes, wrist, or other small anatomic areas. The availability of the light beam permits this severe **primary ray** restriction. This type of tight beam restriction can influence overall film density when one is radiographing the distal extremities with moderate-speed intensifying screens and films in a low to moderate kilovoltage range.

The use of a tightly restricted beam in conjunction with a light beam collimator in Figure C can be compared with the use of the long thin extension cone shown in Figure D. It would be almost impossible to radiograph a distal extremity with the type of cone shown in D because of the lack of light beam feature. The tendency to overcollimate can be almost as damaging as the tendency to undercollimate when a light beam shuttering device is used.

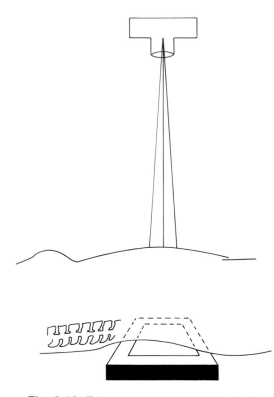

Fig. 2-10. Entrance versus exit beam size. The rectangular light beam pattern that is shown on the surface of the patient is always smaller than the final port size. Minor entrance beam adjustments result in major exit beam changes. It is important that one does not enlarge the entrance beam in an attempt to avoid a cone "cut-off," particularly with a large patient. *Scattered* radiation can increase, significantly destroying radiographic detail.

---

contrast, could conceivably still be under-exposed.

It is difficult to determine whether there is proper beam limiting unless a cone cut is visible. Unfortunately, the opening of a port slightly beyond the cassette size eliminates a visual outline of the shutters. Image quality definitely suffers with the larger beam sizes, but it is difficult to prove to an adamant technologist that he did not use the collimator in a proper manner.

Even when one is properly concerned about limiting the primary ray, there is still a tendency to increase the light beam size

to slightly larger than that required. Take, for example, radiography of the lateral lumbar spine (spot film of L-5, S-1), using an 8 × 10 inch cassette, with a final exiting field size of 6 × 8 inches (48 square inches). Eighty square inches of film are available, but we have elected to use 48 square inches for a good-quality spot film (Fig. 2-10). If it is assumed that the patient is quite large, measuring 30 cm in the lateral position, the light beam diaphragm (at a 40 inch focus-film distance) would be opened to an extremely small beam entrance size, approximately $4\frac{1}{2} \times 5\frac{3}{4}$ inches ($25\frac{7}{8}$ square inches). The rectangular light pattern visible on the patient's skin is rather difficult to accept as adequate, particularly in view of the size of the patient. A tendency of the technologist is to enlarge this opening slightly, to perhaps $5\frac{1}{2} \times 6\frac{3}{4}$ inches, with a resulting exit field of $37\frac{1}{8}$ square inches. Note the *significant enlargement in final field size* as a result of this seemingly small 1 × 1 inch adjustment at the port of entry of the primary ray. This minor adjustment in the shutters in an attempt to salvage a study when one is concerned about making a positioning error can completely destroy image detail, density, and contrast.

Some collimators have a tiny needle-like device attached to the final level of lead shutters (Fig. 2-6). When the collimator is restricted to the proper field size, these tiny metallic structures are projected on the lateral aspects of the finished radiograph (Fig 6-4). Since these needles or arrows are close to the source of radiation, they do not stay in focus and are demonstrated as fuzzy penumbra-like densities. These metallic markers are occasionally placed on the final shutters so that they are in alignment with the cross-hair centering lines of the collimator. The shadow of the needle is superimposed on the shadow of the cross-hair and is not seen in the illuminated field by the technologist. Small pieces of metal attached to the shutters of existing collimators serve the same purpose, helping to improve radiographic quality. If the penumbra-like hazy densities of the metal markers are not visible on the lateral aspects of the radiograph, the shutters were opened beyond the required field size.

Radiation Protection Advantages Using Automatic Collimation Devices

A principal cause of unnecessary exposure to patients is the use of an excessive beam size during conventional diagnostic procedures. A recent survey conducted in the New York area demonstrated that 50 per cent of the beam sizes used for radiography of the chest exceeded 33 inches in size.[9] With automatic collimation this obviously could not occur. Although the automatic collimator does not eliminate gonadal exposure if the gonadal area is in the primary beam, significant dosages to the gonadal area are reduced by the use of the automatic collimator when other sections of the body are examined. When the field size is larger than the film size, the gonadal dosages are from 80 to 400 times higher. Proper primary ray collimation can be extremely effective in lowering patient dosage, particularly gonadal dosage.[7]

GRIDS, MOVING OR STATIONARY

Dr. Hollis E. Potter in one of his earlier papers in 1920, outlined the basic principles of the device which now carries his name, and that of Dr. Gustav Bucky.[14] Dr. Potter fashioned a grid similar to the original Bucky grid, except that the lead lines, which are parallel in nature, were made slightly concave to overcome centering and focusing difficulties. By means of a simple oil-drag system he caused the grid to travel a few inches during the radiographic exposure. This cross-table motion of the grid resulted in a blurring of the grid lines parallel to the length of the table. By blurring the grid lines the quality of the radiographic image was significantly improved, as compared with earlier films using a stationary grid. Although a common term for the Potter-Bucky diaphragm is the word "Bucky," the stationary grid was invented by Dr. Bucky[1] in 1913 and the principle of moving the grid to obliterate its lines was contributed by Dr. Potter.[3]

How many different types of grids should be available to the average radiologic department? A typical reaction to the question, "Would you like a specific grid for a specific purpose?" is the response: "We have a grid." The grid in question can be anywhere from a year to 40 years of age, may have 60, 80, or more, lines per inch, can be focused at 30, 40, 60, or 72 inches, and yet is used for every type of grid study in or out of the x-ray department. A chest grid focused at 72 inches should not be used a standard 40 inch grid, and 40 inch grids should not be used at 72 inches. Very few departments have a complete selection of grids.

Grids are not expensive. They may seem costly as an initial investment, but they are not expensive, for they increase the versatility of the equipment. A 15 × 18 inch high ratio grid for chest or table radiography can be purchased for less than $300. Many grids retail at a more reasonable rate, and yet modern departments will purchase a $10,000 or $15,000 mobile radiographic unit and continue to use an old grid in conjunction with this modern piece of equipment. Even more distressing is the fact that as bedside radiographic quality increases, requests for bedside examinations increase, and the grid is kept out of the department for longer periods of time. The departmental grid then becomes the mobile unit grid.

It would be helpful if an unlimited number of grid cassettes were available for bedside work, unusual examinations, operating room technics, etc., but grid cassettes are significantly more expensive than conventional cassettes. Grid frame devices can be snapped onto an existing cassette to hold a grid in proper position. These snap-on frames make available an unlimited number of cassettes for grid technics.

Grid Interchangeability

An ideal situation would be to have a 2nd grid that would fit into the typical Potter-Bucky diaphragm housing. Most American-made Potter-Bucky housings are fitted with standard-size interchangeable grids, and a typical Bucky grid is very easy to remove or replace if simple directions are followed. First, remove the Bucky tray. Second, never completely remove the grid support rack, for it is difficult to replace. Third, when the grid housing rack is partially removed, simply lift out the grid. A grid storage rack should be provided to house the grid that is not in use. To avoid a grid mix-up, color code the grid with a large-stroke water-soluble felt tip marker. Write directly on the grid its ratio: 8 to 1, 12 to 1, or 16 to 1. *Clearly mark tube*

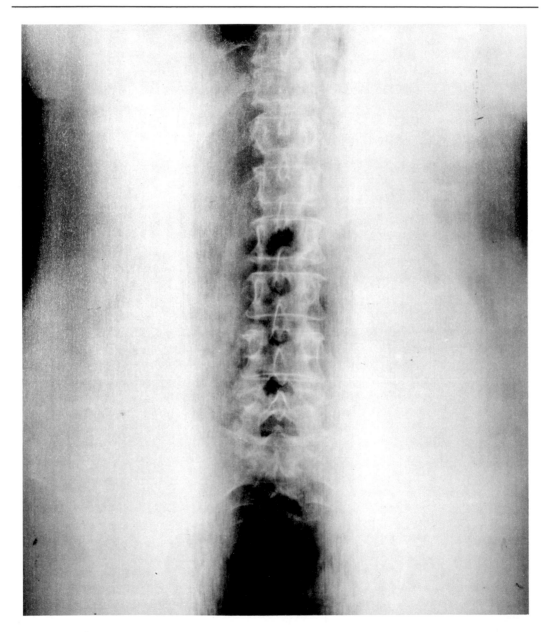

Fig. 2-11. Grid in reverse in Bucky. Shown is the result of placing a high-ratio grid (16 to 1) in the reverse position in a Bucky. The parallel grid lines or center of the grid permit the passage of x-ray in the reverse position.

side to avoid reversing the grid in the Bucky, for if a grid is truly focused, that is, if all of the grid lines converge at a fixed point, the reverse placement of a grid results in technical difficulties (Fig. 2-11).

Why should there be variation in grid ratio or more than one grid available? Grids, whether stationary or moving, have definite kilovoltage limitations, based on a full-size field. For example, an 8 to 1 grid, while adequate for most pediatric or routine radiographic work, is restricted at a full field-size to 95 kVp for effective scattered radiation clean-up. A 12 to 1 grid, while adequate for most barium studies, is limited to 110 kVp and leaves something to be desired in examining large patients, particularly during barium procedures. A 16 to 1 grid is superb for barium enemas, pelvimetries, or obese patients, functioning effectively up to 125 kVp. The difference between an 8 to 1 grid and a 16 to 1 grid enables the technologist to use up to 125 kVp instead of 95 kVp. It seems senseless to buy an x-ray machine with a 125 to 150 kVp output and then be restricted from using these kilovoltage ranges because of the limitations of the grid. It is no more practical to put an 8 to 1 grid in a radiographic table combined with a high kilovoltage generator than it would be sensible to put a 16 to 1 grid in every unit in a radiographic department. It seems more logical that an 8 to 1 or a 12 to 1 grid be used for routine radiography, and that the 16 to 1 grid be available for specific studies of thicker body parts.

Let us compare different grids with regard to technic and the generation of heat units. Assume that a lateral lumbar spine technic requires 90 kVp, 200 mA at 2 seconds with an 8 to 1 grid. The total combination of these factors results in 36,000 heat units. If an adequate x-ray tube were available and were used in conjunction with a 16 to 1 grid, the original mAs value could be lowered to half (200 mA at 1 second). By adjusting the original 90 kVp to 110 kVp, we make several technical gains. First, we improve our technical latitude because of higher kilovoltage.

Second, we shorten our exposure time (actually cutting it in half), and, third, the new technical factor combination results in the production of 22,000 heat units as opposed to the original 36,000 heat unit value, a saving of 14,000 heat units per exposure. *A heavy-duty radiographic tube is a prerequisite for technical adjustments of this type.*

Technical Factor Conversions

There are many rules of thumb on factor conversion for the utilization of a grid. McInnes suggests that the grid gain factor be considered as times 4.[11] For practical purposes an increase of 4 times in the mAs value over the nongrid technic is acceptable.

In converting from a nongrid technic to a grid technic, the addition of kilovoltage rather than the increasing of the milliamperage-second value is advised. If a grid technic must be converted to a nongrid technic, the conversion should be made by the reduction of the milliamperage-second value as opposed to a reduction in kilovoltage values, since adequate penetration must be maintained. This is good practice unless the kilovoltage values are excessively high. Careful attention must be paid to the collimation of the primary beam to help compensate for the lack of a grid.

A basic exposure compensation chart follows to help formulate technics in converting from a nongrid to a grid technic. This chart should be used as a guide rather than a technical guarantee.

Grid Ratio	mAs Increase	kVp Increase
5:1	2×	+ 8kVp
8:1	3-4×	+15 kVp
12:1	5×	+20-25 kVp
16:1	6×	+20-25 kVp

There is some disagreement about when to utilize the grid in conventional radiographic technics. In general, if the area of the body exceeds a measurement of 10 centimeters, it is good technical policy to utilize a grid or Bucky. This technical generalization can be ignored for certain studies, the chest being the most important example. Many departments use a nongrid technic for radiography of the chest due to the radiolucent nature of the lungs. Although a chest will measure in the same centimeter range as the abdomen, less scattered radiation is generated as compared with the abdomen or other dense portions of the body. Areas such as the knee or shoulder can be examined without the use of a grid or Bucky if maximum restriction of the primary ray is accomplished.

TYPES OF X-RAY GRIDS

Stationary Grids

The most popular stationary grids vary from 60 to 80 lead lines per inch, with a low density interspacing material.

Parallel Grid. One type of grid in common use is called a parallel grid. The lead strips of a parallel grid are not focused to a predetermined point, but are positioned vertically across the entire grid. These grids are generally low in ratio, and the thickness of the grid can decrease toward the sides of the grid, with grid ratio decreasing on its lateral aspects. This "prismatic" section grid, in which the heights of the unfocused lead lines gradually taper towards the edges of the grid, is used to overcome possible variations in focus-film distance without causing significant grid cut-off. Parallel grids are generally used with technics under 100 kVp in conjunction with bedside or mobile technics. They are occasionally used in the operating room, where positioning is difficult. Parallel grids that are not prismatic should not be used under a 48 inch focus-film distance unless they are used with an 8 × 10 inch cassette or smaller.

Focused Grid. The second major type of stationary grid is the focused grid. The lead strips are positioned in such a way that all sides of the strips are centered toward a single centering point at a predetermined distance (Fig. 2-12A and 2-12B). For example, if a grid is said to be focused at a 40 inch focus-film distance, one could draw a line

from the most lateral single grid line in a full width grid, and this line would intersect at a given point—in this case at 40 inches—with a line drawn from the most lateral single grid line on the opposite side of the grid. Focused grids are critical in regard to variations in focus-film distance, and accurate tube centering is a *must* (Fig. 2-13).

Basic Grid Design. Regardless of the type of grid used, basic grid design is the same. Lead grid lines are used to absorb scattered radiation. These lead strips are made as thin as possible and separated from one another by a radiolucent material such as paper, plastic, or aluminum. The lead strips are generally about 0.05 mm thick, and they are held roughly apart from each other by 0.33 mm radiolucent spacers.[16]

Grid ratio is defined as the ratio of the height of the lead strips to the distance between each strip. As grid ratio increases, the height of the grid lines increases, but the distance between the lead strips remains the same (Fig. 2-12A).

A frequent description applied to grid evaluation is the number of lead lines per inch. Grid lines vary from a minimum of 60 lines per inch to a maximum of 133 lines per inch (Fig. 2-14A). Most American-made grids have 80 lines per inch (Fig. 2-14B). The 80 line per inch grid contains approximately 1,120 lines plus 1,120 interspaces over the 14 inch width of the grid.

Cross-Hatch Grids. Two stationary grids of any type that are used one on top of the other, each at a right angle to the other, result in a cross-hatch grid (Fig. 2-15). Whether the grids are parallel or focused is immaterial. If the grids used are of different focus film lengths, the technologist must be concerned about the focal range of the most critical grid. For example, if an 8 to 1 focused grid with a focus-film distance of 34 to 44 inches is used in combination with a 5 to 1 focused grid with a focus-film distance of 28 to 72 inches, the cross-hatch combination must be considered an 8 to 1 focused grid. Cross-hatch grids should be used only when very high ratios are desired. A distinct disad-

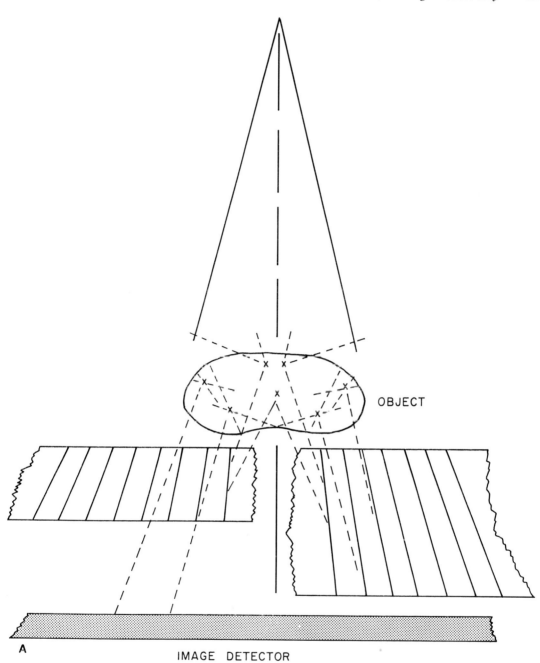

A

IMAGE DETECTOR

Fig. 2-12A. Grid ratio. The section of grid on the left is of a lower ratio than the section of grid shown on the right. *Scattered* radiation that is able to bypass the lower-ratio grid is absorbed by the higher-ratio grid.

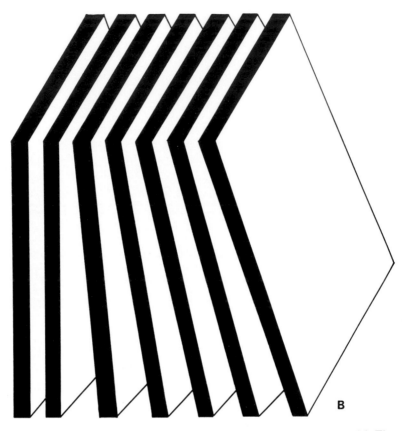

B

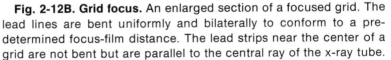

Fig. 2-12B. Grid focus. An enlarged section of a focused grid. The lead lines are bent uniformly and bilaterally to conform to a predetermined focus-film distance. The lead strips near the center of a grid are not bent but are parallel to the central ray of the x-ray tube.

vantage of the cross-hatch grid is the necessity of centering the x-ray beam to the center of both grids. *Not only must the central ray be centered to both grids, but it must also always be perpendicular to the film plane.* Although the original grid used by Bucky was cross-hatch in nature, only recently have cross-hatch grids been accepted.[1]

The use of a cross-hatch grid is quite common for angiographic technics except for cerebral angiography in which tube angulation is required for the Towne projection. These grids were popularized by angiographic technologists, who have to deal with the increased amounts of scattered radiation generated during simultaneous biplane studies. Other uses for cross-hatch grid combinations include transtable lateral projections during myelography in which high kilovoltage can be used (Fig. 2-16), or lateral hip radiography in which high quality radiographs of the head of the femur and its relationship to the acetabulum are required.

A simple way to fashion a cross-hatch grid for the lateral hip or myelogram technic is to use a 10 × 12 inch grid cassette and place a 5 to 1 grid in front of it. The central ray must remain perpendicular to both grids. *The grid lines of the superimposed grid must run opposite to the grid lines of the grid cassette.*

A common error is the placing of a grid over a grid cassette with overlapping of the grid lines of both cassettes. Since it is virtually impossible to superimpose both sets of grid lines, one on top of the other—particularly if the grids are manufactured with different lines per inch design—the resulting superimposition or approximate superimposition of grid lines creates a "moire" pattern (Fig. 2-17). A simple way of demonstrating this "moire" pattern is to radiograph 2 stationary grids at a low mAs, moderate kVp range (approximately 45–50 kVp), so that the grid patterns can be seen on the finished radiographs. Place one film on top of the other on an illuminated view box with the grid lines running in the same direction. Move one of the films 1 or 2 degrees in either direction to create an endless variety of "moire" patterns. This pattern is occasionally encountered where stationary grids rather than moving grids are used for chest x-rays. If a 14 × 17 inch grid cassette is inadvertently used with a stationary grid, the "moire" pattern appears. Likewise, if a stationary grid is installed in an x-ray table in place of a moving Bucky, the use of a grid cassette in the table tray with the grid lines running in the same direction as the stationary grid results in a "moire" artifact. Generally, these films are grossly underexposed as the superimposition of lead lines produces an almost solid lead wall. If this technical accident occurs in using a phototimer, the phototimer attempts to compensate for the lower density of the film. In either instance the radiograph is not diagnostic.

If a high-ratio grid (12 to 1 or 16 to 1) is not available and an occasional large body area must be radiographed—for example, a 30 to 35 cm lateral view of the lumbosacral area—it is possible to make a cross-hatch combination using an existing grid cassette in a transverse position in the Bucky tray. The patient is positioned for the lateral spot film, the primary ray is restricted to proper field size, and a film is made. When the grid cassette is placed in a transverse position, its lines run at a direct right angle to the grid lines of the table grid or Bucky (Fig. 2-18). If a 5 to 1 grid is used in conjunction with an 8 to 1 Bucky, the combination results in a cross-hatch clean-up capability of approxi-

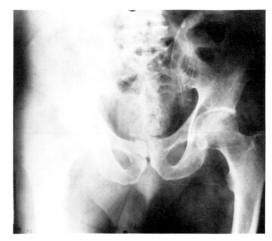

Fig. 2-13. Tube off-center in relationship to Bucky. Note the severe differences in radiographic density from the right to the left side of the pelvic film.

The right side of the radiograph seems adequately exposed, while the left side is significantly less dense. This radiograph was made with a 16:1 Bucky, and the radiographic tube was positioned approximately 1½ inches off center to the left side of the pelvis. Higher-ratio grids are more subject to this type of cut-off than are lower-ratio grids.

This defect is a common happening in using a 12:1 ratio or 16:1 ratio grid. The central ray strikes the slanted or focused lead lines of the grid, which act as an almost solid lead barrier. The focused lines on the opposite side of the grid (*right*) permit the passage of a higher percentage of the lateral aspects of the beam.

mately a 13 to 1 grid. This technic is particularly helpful in a small department in which only one radiographic table with a low or moderate ratio grid exists.

Types of Grid Motion in a Potter-Bucky Diaphragm

The purpose of moving a grid during an x-ray exposure is to erase the grid lines from the finished radiograph. The grid is moved a short distance at right angles to the grid lines (Fig. 2-15). To avoid a grid focusing problem, the lead strips must not be moved

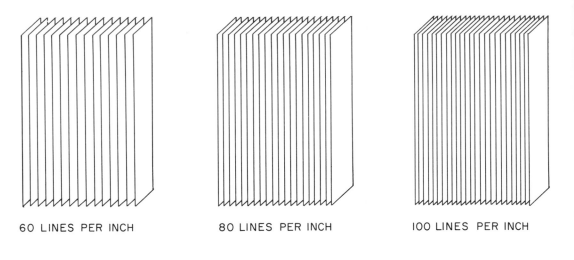

60 LINES PER INCH   80 LINES PER INCH   100 LINES PER INCH

A   SCALE: 2" = 1/4"

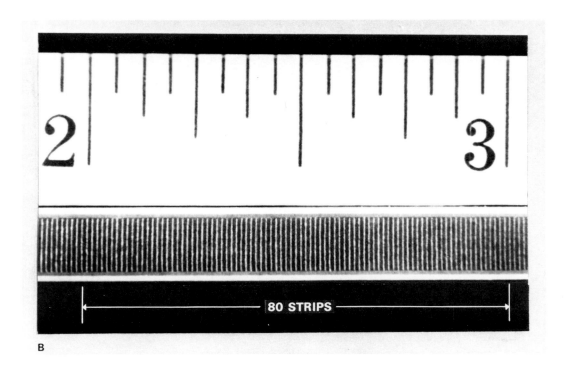

B

Fig. 2-14. (A) Lead lines per inch. Grids are available with several lines per inch; totals of 60, 80, and 100 lines per inch are shown. As the lines-per-inch total increases, manufacturing difficulties increase. Most American-made grids have 80 lines per inch.

(B) Enlarged photograph of a typical 80 line per inch grid. Note 80 strips of lead (white lines) separated by 80 radiolucent interspaces (black lines) in a 1 inch space. (Liebel-Flarsheim Co., Cincinnati, Ohio)

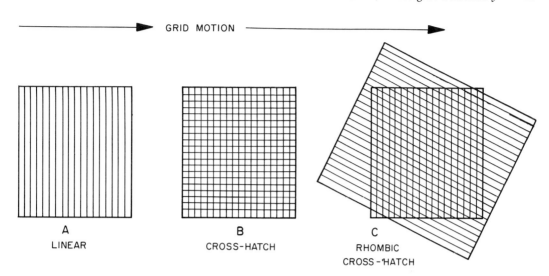

Fig. 2-15. Bucky movement and grid patterns. The grid pattern of a conventional linear grid is shown (A), and it is easy to see that if the grid were to be moved in the direction of the accompanying arrow, the grid lines would be erased or blurred out during a typical exposure.

The next grid pattern represents a cross hatch grid combination (B). The individual grids are placed at a right angle to each other. This cross-hatch effect has several disadvantages, the major disadvantage being that tube-angled technics cannot be utilized with this type of device. Another disadvantage is that a cross-hatch grid cannot be utilized in a Bucky mechanism, for if a grid of this type were used, the lengthwise grid lines would be obliterated, but the horizontal or transverse grid lines would not blur out because they would be moving in the same direction as the grid.

The third drawing represents a compromise (C), the utilization of a cross-hatch type of grid in conjunction with a moving Bucky mechanism. By using a rhomboid pattern, good grid obliteration can be achieved.

The linear lines of the grid run in the same direction as those of a conventional grid. The transverse lines are placed obliquely or tangentially to the linear lines. The angulation of the overlay grid permits obscuring of the lines in the Bucky mechanism.

A linear grid larger than that needed is used to demonstrate how the rhombic effect is achieved (C).

Cross-hatch grids or rhombic grids are generally sealed under a common aluminum covering. Although the rhombic pattern grid has never achieved popularity, the cross-hatch technic has been widely accepted, particularly for angiography.

---

over an exaggerated area (several inches or more). If a grid is moved over too short an area, there will be an imprint of the grid lines on the finished film. As the manufacturing of grids has improved and lead lines have become finer, the distance over which the grid must be moved for sufficient blurring has been decreased.

Single Stroke Motion. Several types of grid motions are available, including single stroke motion. The Bucky diaphragm is cocked by means of a spring mechanism, and the spring is used to move the grid at a predetermined uniform speed. An oil-dash pot can be adjusted to control the speed of the movement of the grid. The motion is about one half inch to either side of the center of the grid.

Reciprocating Bucky Diaphragm. A repeated to-and-fro motion is used with a reciprocating grid. The grid reaches its maximum speed quickly, an oil-dash pot con-

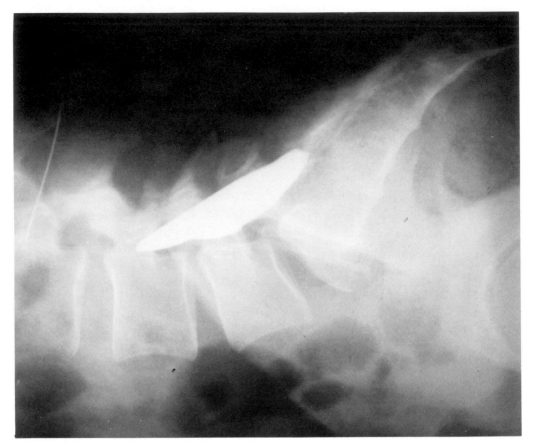

Fig. 2-16. Use of a cross-hatch grid. A cross-table lateral projection during myelography can be made by using 2 linear grids as a single cross-hatch grid. Two 6 to 1 ratio aluminum interspaced linear grids were used to make this study at 120 kVp. A grid cassette can also be used with a single linear grid.

trolling its speed. Reciprocating grids are available in conventional and high-speed models. The conventional model is used with a 60 line or an 80 line per inch grid and has an exposure tolerance as low as $\frac{1}{10}$ of a second. The superspeed model, which moves considerably faster, is used when obliteration of grid lines, generally 80 lines per inch, is necessary with exposures up to $\frac{1}{60}$ of a second. Faster exposure times can be used, but grid-caused density variations can appear.

Rhombic Pattern in a Cross-Hatch Grid. A cross-hatch grid using a rhombic pattern can be used with a superspeed reciprocating Bucky, particularly with kilovoltage values in excess of 100 kVp. Because of its cross-hatch effect, tube-angle technics cannot be used

with the rhombic Potter-Bucky diaphragm (Fig. 2-15). This grid is no longer commercially available.

Circular Motion Grids. Special purpose grids which rotate in a circular fashion are available for cassette or film changers. Because these grids rotate at extremely high speed, they have excellent grid line erasure. Tube-angle technics are prohibited with this system.

OTHER ASPECTS OF GRIDS

Radiographic Properties of Interspacing Materials

Basic interspacing materials are of 2 types, organic or inorganic. The organic material, paper, plastic, cardboard, etc., has a very low absorbing property over the entire range of diagnostic x-ray. The inorganic spacing material, aluminum, has a higher absorbing property. Aluminum readily absorbs scattered radiation as well as a small percentage of the primary beam. Density variations, which can be caused by grids with inorganic interspacing, do not generally occur with aluminum interspacing.

A corduroy pattern can occasionally be seen on a Bucky radiograph using an inorganic interspaced grid. This pattern is frequently seen on angulation technics such as the tunnel view of the knee. A slight medial or lateral tilting of an angled tube can produce this effect (Fig. 2-19). This striped pattern can also occur when a focus-film distance is used that does not coincide with the recommended focus-film distance of the grid in use. As grid ratio increases, the problem is more likely to occur.

Aluminum Interspacing. Since grids should offer a minimum of resistance to the primary beam, there is some disagreement about the use of aluminum as an interspacing material as opposed to the use of organic materials. While it is true that aluminum interspacing absorbs some of the primary ray, it also serves a valuable function (Fig. 2-20). Scattered radiation from the lead lines themselves is absorbed by the aluminum interspacing material. In effect, the inorganic (aluminum) interspacing of an all-metal grid becomes the "grid's grid." In a direct comparison of existing grids of the same ratio but of different interspacing materials (inorganic versus organic), definite improvement in the radiographic image secured with the all-metal grid is obvious. The presence of the aluminum interspacing acts as an added reducer of scattered radiation, thereby improving image quality.

Grid Manufacturing Difficulties

Canting. The process of manufacturing a grid and inclining its lead lines to a fixed focus film distance (canting) is extremely slow and difficult. Because of the canting process all present-day grids can be made

Fig. 2-17. "Moire" grid artifact. This "moire" or "zebra" artifact is the result of a common error in attempting to use cross-hatch grid combinations.

When a grid cassette is used with another stationary grid to create a cross-hatch effect, a technologist on occasion will place the overlay grid on top of the grid cassette with the lines of both grids running in the same direction. The grid is then taped to the grid cassette, usually with a slight degree of misalignment. An exposure is made, and since the grid lines from one grid are superimposed on those of the other grid, film darkening occurs only beneath the superimposed radiolucent interspaces. This film represents an attempt to examine the hip in the lateral projection. The same effect is seen when a grid cassette is accidently used with a stationary grid, whether it be installed in a table or chest changer.

This effect is not noted when a grid cassette is used with a moving Bucky. The moire pattern is common to all double grid technics, but not encountered with Bucky and grid misalignment.

When a grid cassette is utilized with a Bucky, the films are hopelessly underexposed.

This decrease in radiographic density is also noted with dual grid misalignment.

flat. In early days curved surfaces were made which corresponded to the fixed focus-film distance in use; grids formerly were made with parallel lead lines, and then the grid was bent or curved to focus.

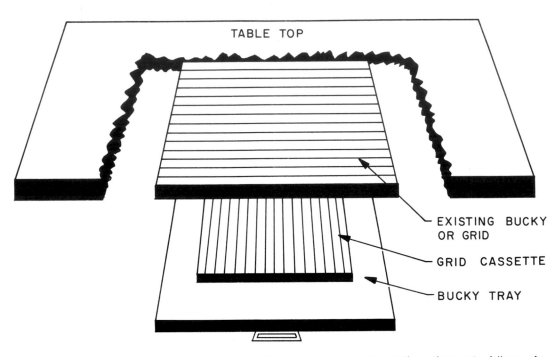

TABLE TOP

EXISTING BUCKY OR GRID

GRID CASSETTE

BUCKY TRAY

Fig. 2-18. "Makeshift" cross-hatch grid device. If a high-ratio grid is not available and an occasional large body area must be radiographed, it is possible to make a sort of cross-hatch combination, utilizing an existing Bucky and a grid cassette. The existing Bucky lines will be blurred out by normal grid motion, but the grid lines of the grid cassette will be seen on the finished radiograph.

If this technic is used with an existing grid rather than Bucky, the grid lines of the table will virtually obscure the grid lines of the grid cassette, and vice versa. One set of lines almost weaves out the other set of lines. An appropriate increase in technical factors must be made. Caution: it is wise to make this increase by means of kilovoltage rather than milliamperage or time. The increase in kilovoltage must be, of course, within the limits of the tube that one is using. If this technic is being used with a phototiming device, it is important to ascertain that the grid cassette is a phototiming cassette. Very few grid cassettes have thinner lead backs, for they are generally basic radiographic cassettes, not phototiming cassettes.

It is apparently easier to manufacture a grid with a more uniform or homogeneous pattern utilizing an aluminum interspacing material. Nevertheless, it is extremely difficult to manufacture any type of grid that is so accurately focused that the lead lines correspond to typical textbook drawings. During the canting process of a grid, the strips of lead are bilaterally bent to focus at a fixed point. At this stage of manufacture, strip shadows can broaden and create uneven radiolucent gaps between the lead lines. Broadening of the lead lines decreases the passage of the primary ray in that area, and the separated or lucent opening allows overexposed linear streaking of the film. The simplest solution for eliminating grid shadows or uneven density variations is to move the grid during the exposure.

Ultimate Goal. Much has been done over recent years to improve the drive mechanism of the Potter-Bucky diaphragm, but the ultimate goal in the making of a grid would be the production of a very fine line grid of such nature that the lines would not be obviously visible to the human eye. The aluminum interspaced grid has virtually reached this goal.

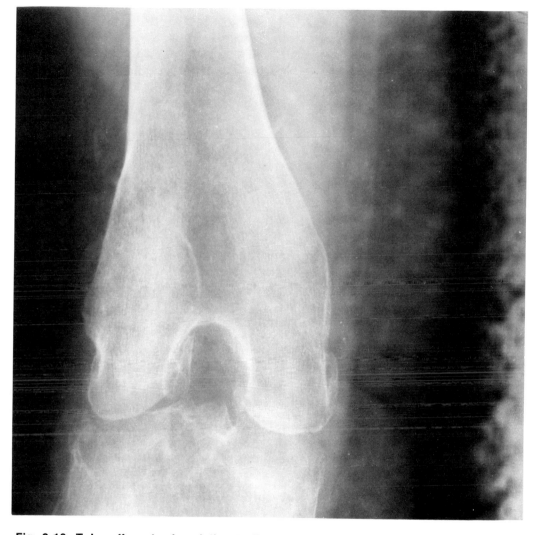

Fig. 2-19. Tube off center in relation to Bucky. A corduroy pattern can develop when a tube is improperly centered or angled to a Bucky. The radiolucent density is the result of accentuating grid impurities by grid motion. The pattern occurs quite often when extremely short exposure times are used.

Modern radiographic units are available with a significant increase in radiation output capability, making shorter exposure times possible. These shorter exposure times "capture" grids in motion, producing variations in grid density. Grid manufacturing impurities that ordinarily would be obliterated by grid motion during conventional exposures are almost unavoidable with extremely short exposures. It seems questionable that a moving vibrating system, such as a Potter-Bucky diaphragm, should be interposed between the remnant ray and the image detector, since an out-of-phase high-speed Bucky can completely destroy image detail and sharpness. A fixed, nonmoving, fine line grid does not have to be moved during the radiographic exposure, and, if properly manufactured, the grid lines should be virtually invisible and therefore not visually offensive to the radiologist interpreting the films.

While the use of an aluminum or inorganic

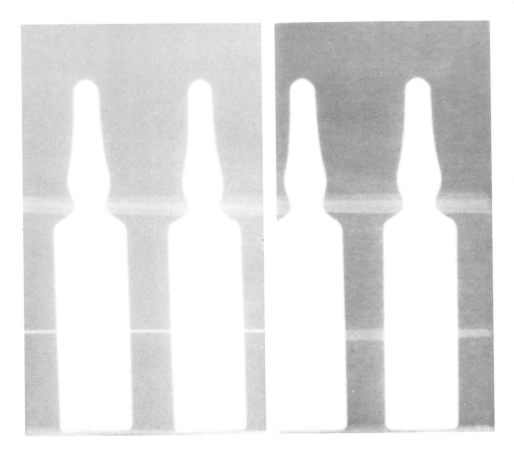

Fig. 2-20. Inorganic versus organic grid interspacing material. Inorganic (aluminum) interspacing material is slightly more restrictive to the primary beam than is organic interspacing material. Both radiographs were made with a single exposure. Both grids were at the same focus-film length and were 10 to 1 in ratio. The grid on the right uses aluminum interspacing material. Primary beam attenuation is barely noticeable.

spacing material seems relatively new, Dr. Bucky as early as 1915 or 1916 had proposed an outline for a grid of this type. These grids were described in the patent literature of that time, but because of production difficulties never reached the manufacturing stage.[10]

Preventing Grid Impurities. Since the heart of the problem of grid impurity or density variations is the manufacturing process, one should radiograph every new grid for uneven density patterns. *Do not accept a grid for an old or a new installation unless you see or take an industrial radiograph of that specific grid.* A low exposure (5 to 10 mAs in the 45 to 50 kVp range) usually produces a satisfactory industrial radiograph of a grid.

Although some grid impurities can be "erased" by the moving of a grid, they cannot be avoided with stationary grids. Dropping or damaging a grid will tend to produce or accentuate radiolucent defects. The use of a "fractured" linear grid puts an intolerable strain on the physician required to interpret a radiograph. Occasionally an x-ray department will have 2 grids with severe fractures, and yet when these grids are used in combination as a cross-hatch grid, one grid tends to obliterate the impurities of the other.

Unwanted Grid Lines on Radiographs

There are many causes for unwanted grid lines on a radiographic film. First, the most

Fig. 2-21. Grid focus. Improper centering of an x-ray tube to a grid or Bucky is one of the most common causes of unwanted grid lines on a radiograph.

The top frame was made with the x-ray tube centered to a 16 to 1 Bucky. There is an overall evenness of density to the radiograph.

The middle frame was made with the tube off center (1 inch) to the right. The decrease in radiographic density on the right side is noticeable.

The lower frame was made with the tube off center (3 inches), also to the right of the center of the Bucky. On the right side there is a complete absence of the image. On the left side of the film there is also a decrease in radiographic density.

All 3 exposures were made with the same cassette, a single sheet of 14 x 17 inch film, and identical technical factors.

common cause is the improper centering of the x-ray tube to the grid. This problem becomes more critical as the ratio of the grid increases. An 8 to 1 grid is not nearly so critical from a centering standpoint as is a 16 to 1 grid (Fig. 2-21). Second, every grid has a predetermined focus, and if there is a significant increase or decrease in focus-film distance, a widening of the grid lines on the lateral aspects of the film is apparent, with a decrease in radiographic density. This can occur with a stationary grid or a Potter-Bucky diaphragm. Third, if an exposure time is entirely too short for a moving grid, the grid is "captured" in motion. Fourth, the x-ray exposure may be started before the grid is put in motion or continued after the grid has ceased to move. Fifth, a stereo technic may be used, with the tube shifting opposite to the direction of the lead strips.

OTHER TECHNICS AND DEVICES

Scattered Radiation Reduction in Hazardous Areas

It is not safe to use a light beam collimator in the operating room or other hazardous areas during certain procedures. The light beam mechanism in the collimator is not hazard-proof and should not be used when an explosive gas may be in use.[19]

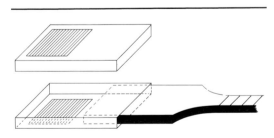

Fig. 2-22. Operating room cassette tunnel. A grid of good quality can be laminated between 2 layers of plywood on one side of a cassette tunnel. A tray with a long handle can be used to position a cassette within the tunnel. The open end of the tunnel is placed opposite the surgical field.

By turning the tunnel over, nongrid films can be made.

It is impractical to use a Bucky in an operating room for several reasons. First, the Bucky can be bulky in size and hard to position under a patient. Second, it can be troublesome to operate, particularly if a hand-cocked Bucky is used and must be hand-triggered. The use of an electrically operated Bucky produces an intolerable hazard with certain types of anesthesia. The potential of an explosion is too great to risk; therefore the use of a stationary grid with organic or inorganic interspacing is recommended for operating room radiography.

A simple device can be constructed for use in the operating room for conventional grid radiography (cholangiography, hip nailings, etc.). This unit consists of a plywood or conventional bakelite film tunnel. Grid cassettes can be placed in position manually or with a long tray-like attachment.[4] It is recommended that a single good-quality grid be purchased and laminated between 2 thin layers of plywood or bakelite on one side of the cassette tunnel. A built-in grid would then be available. Should a tunnel be required for a nongrid procedure, reversing the grid tunnel would be adequate (Fig. 2-22). Another grid could be kept in a snap-on device to be used with an unlimited number of conventional cassettes for lateral or cross-table projections.

A difficulty encountered *during operative cholangiography* is the superimposition of the biliary ductal system upon the spine on a finished radiograph. It has been recommended by Custer and Clore that the operating table be rotated 15 degrees to the right to cast the shadow of the biliary system away from the spine. When a patient is positioned in the 15 degree right posterior oblique projection, the biliary system, being anterior, will rotate laterally, and the lumbar spine will rotate medially.[6] A more simple technic to avoid superimposition of the biliary tree on the spine would be angling the x-ray tube 20 degrees from left to right. The x-ray machine is placed on the left side of the patient, and the x-ray tube is angled 20 degrees to the right side, with the central ray entering

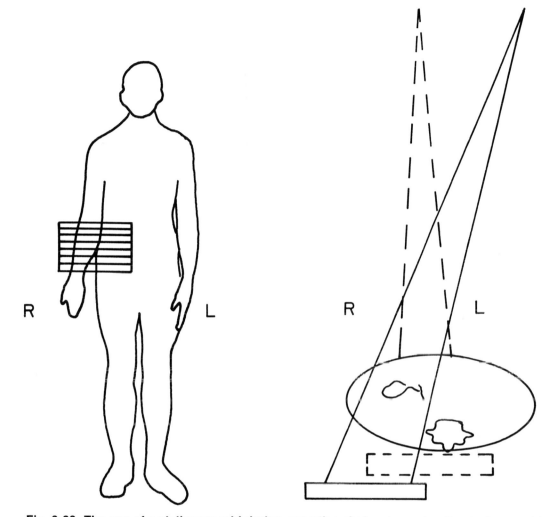

Fig. 2-23. The use of a stationary grid during operative cholangiography. By placing a grid or grid cassette in a transverse position during an operative cholangiographic study, an "oblique" radiograph can be made of the biliary ductal system. The common duct can be projected off the spine without the rotation of the patient.

An added advantage of this technic is that both tube and film can be positioned on the left side of the patient, opposite to the operating field, helping to reduce the possibility of contamination of the sterile field.

---

the area of the biliary ductal system (Fig. 2-23). Because the biliary ductal system is anterior to the spine, the biliary ducts will be projected to the lateral aspect of the patient's abdomen. The spine, being posterior, shifts very little on the finished radiograph. The cassette must be centered slightly lateral to the right side to allow for this tube tilt technic. *The grid cassette or grid tunnel must be placed*

in a transverse position; that is, the length of the cassette or grid must run with the width of the table rather than with the length of the table.

The Use of a Grid in a Spot Film Device

Most fluoroscopic spot film devices contain a 5 to 1 or 6 to 1 ratio grid although some

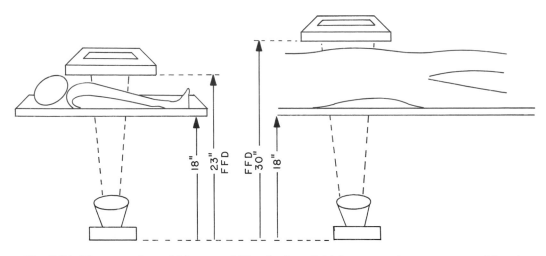

Fig. 2-24. The use of a grid in a spot film device. Grid focus can be a severe problem in a spot film device because of the variability of the focal object distance. A low-ratio grid with a generous focus-film length is usually used.

are available with 8 to 1 focused grids. Since there is a fixed distance from the tube to tabletop, the spot film housing varies in object-film distance, constantly changing the focus-film distance. For example, an infant in the AP position might require the placement of the spot film tunnel 3 inches from the tabletop for a 21 inch focus-film distance. The same fluoroscopic spot device in use with an adult in the lateral position might require the fluoroscope to be 12 inches away from the tabletop, with a 30 inch focus-film distance (Fig. 2-24).

Many spot film devices have oscillating or vibrating grids, and some fluoroscopists feel that this is a *must* for spot films of the gallbladder, because the motion of the grid eliminates grid impurities, enabling the radiologist to visualize small radiolucent defects in the gallbladder. When a moving grid is not available, many radiologists feel so strongly about grid impurities that they remove the stationary grid for horizontal beam spot films of the gallbladder. Figure 2-25A shows how to position a stationary grid when there are focus difficulties.

However, grid lines in virtually every make of x-ray equipment run perpendicular to the floor, exactly opposite to the leveling of the

stones (Fig. 2-25B). It is true that a single stone or a small polypoid-type of lesion in the gallbladder could be lost within the impurities of a stationary grid, but it would be almost impossible to miss a leveling of stones in either the erect or the decubitus position with the grid lines running perpendicular to the floor.

Chest Radiography Using a Grid or Bucky

Wall-mounted cassette holders for chest radiography have been decreasing in size over the past 10 years. In the mid-1950's, bulky cassette changers that incorporated stereo shift mechanisms were quite common. These large-size cassette changers were equipped with a door that contained a Potter-Bucky diaphragm, and were placed sufficiently far from the wall of the x-ray room that a patient could be positioned sitting on a stretcher, with his legs over the side for an upright PA chest radiograph. As fine line grids were accepted and bulky mechanisms were no longer needed to move the grid, inexpensive wall-mounted cassette holders with a stationary grid became available. Many of the large-size cassette changers sold

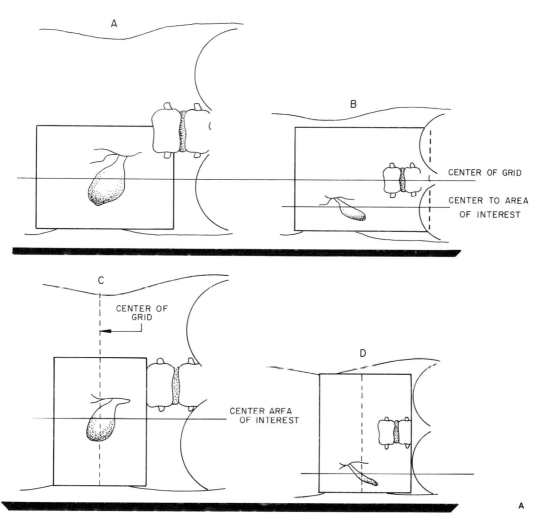

Fig. 2-25A. Stationary grid positioning and focus difficulties. The lengthwise (C and D) rather than the horizontal (A and B) placement of a grid permits the centering of a well collimated beam to the exact area of interest without the risk of grid focus difficulties.

Fig. 2-25B. Horizontal beam gallbladder films. Grid lines should be placed perpendicular to the floor when the gallbladder is investigated in the erect or decubitus position. If a radiolucent defect should exist in a grid, it would then be positioned perpendicular to layered gallbladder calculi.

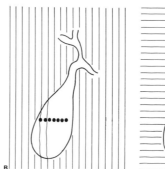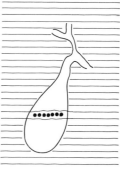

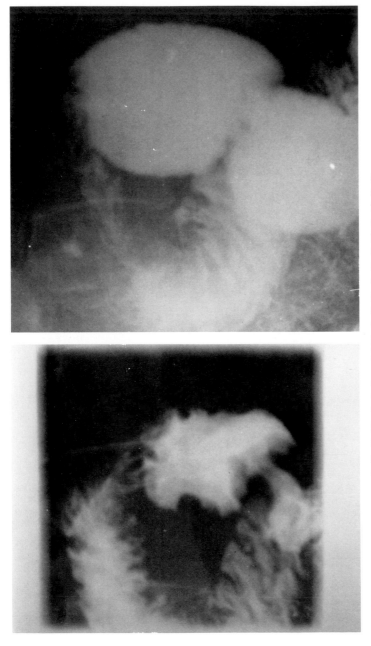

Fig. 2-26. Necessity of restricting primary beam to size of lead mask. Upper frame: a 4 x 5 inch segment of a fluoroscopic spot film of the duodenal bulb and loop made with a full field beam (10'' x 10''). Lower frame: a 4 x 5 inch segment of the same duodenal area (different stage of emptying) made with a well collimated primary beam and a protective lead mask at the film site.

Both exposures were made with the use of 200 mA at 100 kVp for $\frac{1}{10}$ of a second. The increase in density in the upper frame can be attributed to an increase in scattered radiation because of the larger field size.

for thousands of dollars, but the new grid holders were available for under $100 each, making it possible to install a chest film holder with a grid in almost every radiographic room. Many large departments soon found the taking of erect chest radiographs on a litter was restricted to 1 or 2 rooms that still contained the old-style cassette changers. If a thin wall-mounted cassette changer is used in conjunction with a grid, it should be mounted at least 12 to 18 inches away from the wall so that a patient on a litter can be placed in the erect PA position, with his legs under the cassette holder.

Lead Rubber Masking

A lead rubber mask is a preparation of lead oxide or salt powder or other attenuating material uniformly distributed in a rubber or plastic base in the form of flexible sheeting.[17] The use of a lead rubber mask to subdivide cassettes is both esthetic and economic. More elaborate lead masking devices include (1) a fluoro spot film mask to divide spot films in equal halves for such technics as esophageal radiography or myelography; (2) a serial tunnel tray that replaces the Bucky tray in a radiographic table, permitting the division of a 10 × 12 inch or 11 × 14 inch cassette into 4 rectangles of equal size; and (3) a serial tunnel mask which divides an 8 by 10 inch film into 2 circular spots approximately $4\frac{3}{4}$ inches in diameter. A tunnel of this type is frequently used in conjunction with a head unit.

If the primary beam is not restricted to the size of the lead mask, the finished radiograph suffers technically. The difference in proper beam collimation on a serial device versus a full-size beam is obvious (Fig. 2-26). A similar error can be made during fluoroscopic spot filming by pulling a lead mask into place without making a corresponding decrease in the lead shutters beneath the table. With the lead mask divider in position, sharp dividing margins are demonstrated on the finished films. These margins indicate only that the lead mask is in place, for they do nothing to reduce scattered radiation since the primary beam could be open to full port size. This technical error is striking in the making of highly specialized cranial views. *It is mandatory that exact primary beam limiting be used with an exit lead mask.*

The Use of a Lead Rubber Diaphragm at Tabletop

Lead rubber sheeting can be used to restrict or at least effectively to attenuate primary radiation for conventional radiographic studies. There are occasions when, despite careful collimation, severe primary radiation leakage occurs because of the configuration of the

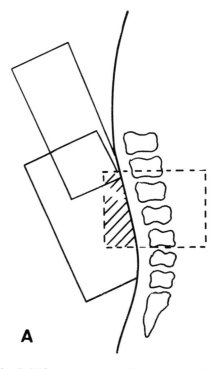

A

Fig. 2-27A. Lead rubber diaphragm at tabletop. When a beam size does not correspond to the contour of the part under study, lead rubber sheeting can be used to shield the cassette in the Bucky tray from *primary* radiation.

part being examined (Fig. 2-27A and B). By placing sheets of lead rubber or lead dividers on the table and carefully moving these dividers into position so that they absorb as much of the unattenuated primary beam as possible, a significant improvement in image quality results.

Compression Devices

Compression devices, such as compression table bands, significantly reduce tissue volume and are helpful in controlling scattered radiation. An abdomen can be compressed several centimeters, and kilovoltage can be appropriately lowered for the now thinner body part, with a corresponding reduction in scattered radiation (Fig. 2-28). Compression of the patient against an erect table or decubitus Bucky reduces tissue volume and

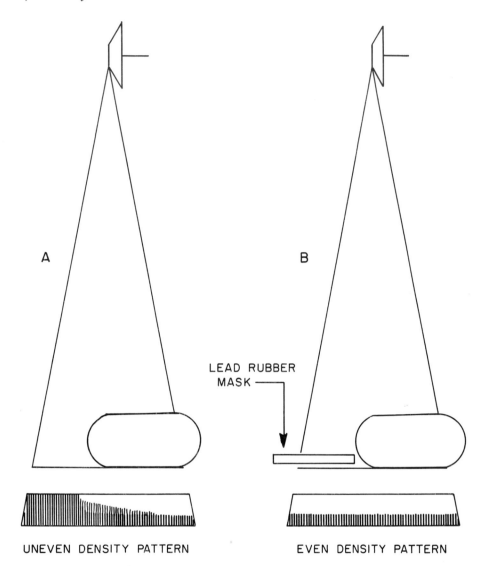

A   B

LEAD RUBBER
MASK

B   UNEVEN DENSITY PATTERN   EVEN DENSITY PATTERN

Fig. 2-27B. When an unattenuated *primary* beam strikes a film, an uneven density pattern of *scattered* radiation can damage film quality. (A) The use of a lead rubber mask at object site (B) can attenuate the *primary* beam, helping to produce a more even density pattern across the entire film.

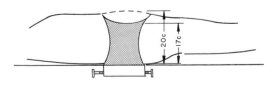

Fig. 2-28. Compression device. The use of a compression device can reduce tissue volume. If a lower kilovoltage value can be used, a corresponding reduction in *scattered* radiation should occur. Compression is most helpful in overcoming patient motion.

therefore reduces scattered radiation. *Compression can add value to the finished radiograph by contributing to the arrest of patient motion, whether voluntary or involuntary.* The use of compression devices for patient immobilization is a lost art, a technical aid that goes a long way toward improving films.

IMPORTANCE OF RESTRICTING THE PRIMARY RAY

The great importance of paying maximum attention to the restriction of the primary ray should be restressed. Any method of primary ray restriction, whether it be by single diaphragm, conventional cone, telescopic cone, multileaf collimator, or an external lead rubber mask—or any combination of the above—will effectively reduce the port entry size of a primary beam and lessen scattered radiation.

REFERENCES

1. Bucky, G.: The elimination of secondary rays in the object during radiography. Verhandlungsbericht der Deutschen Rontgengesellschaft, 9:30–32. 1913.
2. Cahoon, J. B.: Formulating X-Ray Techniques, ed. 7, p. 76. College Station, Durham, N.C., Duke University press, 1970.
3. Characteristics and Applications of X-Ray Grids. rev. ed., p. 3. Cincinnati, Liebel-Flarsheim, 1968.
4. Cullinan, J.: Simplified Procedure Introduced for O.R. Hip Pinning. Dupont X-Ray News, No. 10. Wilmington, Del., E. I. du Pont de Nemours, n.d.
5. Cullinan, J. E., Jr.: Fractional focus x-ray tubes. Newer clinical and research applications. Radiol. Techn. 39(6):333–338, 1968.
6. Custer, M. D., Jr., and Clore, J. N., Jr.: Source of error in operative cholangiography. Transactions of the Southern Surgical Association. 80th Annual Meeting, Boca Raton, Florida, December 9 to 11, 1968. vol. LXXX, pp. 206–213. Philadelphia, J. B. Lippincott, 1969.
7. Epp, R. R., Weiss, H., and Laughlin, J. S.: Measurement of bone marrow and gonadal dose from the chest x-ray examination as a function of field size, field alignment, tube kilovoltage and added filtration. Brit. J. Radiol., 34:85–100, February 1961.
8. Grigg, E. R. N.: The Trail of the Invisible Light. p. 416. Springfield, Ill., Charles C Thomas, 1965.
9. Janower, M. L.: Technological Needs for Reduction of Patient Dosage from Diagnostic Radiology. p. 180. Springfield, Ill., Charles C Thomas, 1963.
10. Mattsson, O.: Practical photographic problems in radiography with special reference to high-voltage technique. Acta Radiol. (Stockholm), Suppl. 120:89, 1955.
11. McInnes, J.: The elimination of scattered radiation. Radiography, XXXVI(426):141–142, June 1970.
12. Peyser, L. F.: Diagnostic x-ray beam collimation. Cathode Press, 23(1):38, 1966.
13. *Ibid.*, pp. 36–43.
14. Potter, H. E.: The Bucky diaphragm principle applied to roentgenology. Amer. J. Roentgen., 7:292–295, 1920.
15. Stanton, L.: Basic Medical Radiation Physics. p. 259. New York, Meredith Corporation, Appleton, 1969.
16. *Ibid.*, p. 275.
17. *Ibid.*, p. 602.
18. Ter-Pogossian, M. M.: The Physical Aspects of Diagnostic Radiology. p. 106. New York, Harper and Row, 1967.
19. Vennes, C. H., and Watson, J. C.: Patient Care and Special Procedure in X-Ray Technology. p. 195. St. Louis, C. V. Mosby, 1964.
20. Walche, R., Stewart, H., and Terrill, J.: An automatic x-ray field size limiting system. Radiology, 89:105–109, July 1967.

3. Radiographic Tubes

The intention of the author in this chapter is to reacquaint the user of radiographic tubes with current tube limitations. A review of common technical errors, errors that lead to partial or complete damage to radiographic tubes, will be presented. Newer concepts of radiographic tubes and radiographic filters will be described.

BASIC COMPONENTS OF A RADIOGRAPHIC TUBE

The actual generation of x-radiation requires many complex circuits within an x-ray machine. The x-ray tube is probably the most important link in the equipment chain, for it is the actual source of radiation. The basic components of a radiographic tube consist of 3 major segments. The 1st is the filament, which is a source of electrons; the 2nd is a target, whose surface is bombarded by these "free" electrons; and the 3rd is a highly evacuated glass envelope, which is used to house the filament and the target (Figs. 3-1 and 3-2).

The filament is generally made of a small coil of tungsten wire mounted in a metal shield or "focusing" cup. Free electrons are achieved by heating the filament within the vacuum of the glass envelope of the x-ray tube. As the filament is heated, electrons move at a speed which is controlled by the temperature of the filament. The higher the temperature of the filament, the faster is the acceleration of the electrons. Because of the increase in temperature of the filament, the electrons reach a point in their acceleration at which they "boil off" in the form of an electron cloud. These negatively charged electrons are set in motion in orbit at the

filament of the x-ray tube rather than across the x-ray tube, an effect known as *thermionic emission.*[8]

It is now necessary to wrest these electrons from orbit across the x-ray tube for the purpose of striking the surface of the anode, which is usually made of tungsten. Tungsten has a high melting point, a high thermal conductivity, and a low vapor pressure at high temperatures. The atomic number of tungsten is 74, and its melting point is 3380 degrees centigrade. A positive source must attract the negatively charged electrons from the filament. A kilovoltage value is impressed upon the tube, and the electrons flow across the x-ray tube to the anode. They travel this short distance at extremely high speed and in striking the solid metal target produce x-radiation. This production of x-radiation is fairly inefficient, for less than 1 per cent of the energy utilized is converted into radiation. Much of the remaining energy is converted into heat, which must be absorbed at least temporarily by the target. Only 0.2 per cent of the energy utilized is converted into roentgen rays, and 99.8 per cent of this energy is changed into heat.[9]

The cathode, or negative electrode of the x-ray tube, is a "hot" cathode, meaning that electrons are continually "boiled off" by the heated filament. The temperature of the filament is varied to increase or decrease the quantity of electrons. Electrons are focused as a controlled stream in their passage from the filament across the x-ray tube to the target. This focused stream of electrons strikes an area on the target known as the focal spot (Fig. 3-1).

The smaller the electron stream, the better is the focal spot; that is, the smaller the area

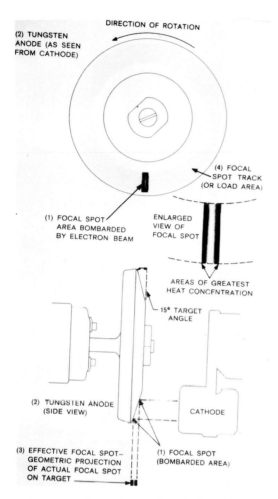

Fig. 3-1. Cathode-anode relationship. The electron beam bombards the area of the tungsten anode (2) known as the focal spot (1). As the anode rotates, fresh unbombarded tungsten (4) is brought into the focal spot area (1). The electron stream strikes the target as 2 almost completely separated energy beams to protect the center of the focal spot from overheating. The focal track (4) is determined by the diameter of the anode disc. The effective focal spot (3) is determined by both the target angle and the filament of the cathode. (Machlett Laboratories, Stamford, Conn.)

difference exists between the anode and the source of the electrons.

If a tube is used improperly, that is, beyond the recommendation of the tube manufacturer, an almost imperceptible pitting of the target occurs, creating tiny depressions in the target area or focal track (Fig. 3-4). These depressions can be invisible to the naked eye but perceptible to the electron beam. Electrons can lose themselves in the depressions, significantly impairing image sharpness as well as radiation output.

All x-ray tubes are enclosed in shock-proof containers. The x-ray tube insert is grounded and is surrounded by oil. Oil insulates the x-ray tube from its metal shield. The x-ray housing which also acts as a mounting for the tube insert, is lined with lead to act as a radiation shielding device (Fig. 3-2). In addition the housing supplies support for the x-ray collimator. An air circulator can be added to the housing to help with the dissipation of heat generated by the use of the tube. An opening is provided in the housing in alignment with the tube window to utilize the most effective portion of the primary x-ray beam.

of the target bombarded, the smaller is the effective focal spot. The angle of the target in its relationship to the electron stream also influences the size of the effective focal spot (Fig. 3-3). The flow of current or passage of electrons occurs only when a high potential

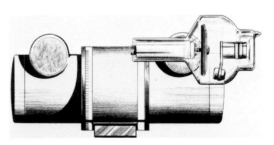

Fig. 3-2. Rotating anode tube insert and housing. The x-ray tube, a highly evacuated glass envelope containing a filament and target, is installed within a shock-proof tube housing. The insert is grounded and is surrounded by oil, which insulates it from the metal housing. The housing is lined with lead to prevent radiation leakage. An opening in the housing permits the use of the most effective portion of the *primary* beam. A cone or collimator is supported by the housing. (Machlett Laboratories, Stamford, Conn.)

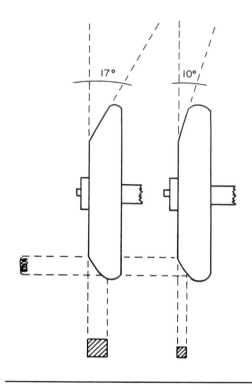

Fig. 3-3. Target angle and effective focal spot size. Steepening of the conventional 17 degree angle target to a 10 degree angle target creates a smaller effective focal spot size. Until recently focal spot size was primarily changed by the reduction in size of the tube filament. With the advent of modern direct roentgen enlargement technics, a way had to be found to permit increased tube loading, for as the tube filament decreases in size, so does the electron stream. Since a smaller electron stream bombards a smaller area of the surface of the target, a reduction in filament size soon reaches a point of no return.

It is interesting to note the differences in the effect of focal spots of the 2 target angles without a change in the size of the filament or electron stream.

Radiographic Tube Filaments

The filaments used in radiographic tubes are relatively small in size. Since filament wire of an extremely small diameter is used, some method of conserving filament life is necessary, for tungsten wire evaporates when heated to very high temperatures. Newer high milliampere output units engender new problems: the higher the temperature of the tungsten filament, the higher is the rate of tungsten evaporation, and the longer the filament is used at the higher temperature, the greater is the amount of tungsten evaporated. When such tungsten evaporation takes place, the diameter of the tungsten filament decreases, limiting its useful life.

TYPES OF RADIOGRAPHIC TUBES

Almost all types of radiographic tubes have 2 filaments which produce a small or large stream of electrons. These filaments are used independently of each other to produce either

a small effective focal spot or a large effective focal spot. *The smaller the effective focal spot, the sharper is the radiographic image.* As the focal spot size decreases, there is definitely an increase in image definition, but also an increase in tube rating restrictions and as larger focal spot sizes are used, there is some decrease in definition. As the focal spot size increases, the filament is correspondingly enlarged, and there is a larger stream of "free" electrons available to bombard the surface of the target. This larger stream of electrons strikes the target, interacting with the dense tungsten over a much greater area. By spreading the electron stream over a larger area of the surface of the target, the technologist is able to use higher milliampere values. Since a larger filament is used for this purpose, it can be safely heated to higher temperatures.

The discussion in this textbook will be confined to the more conventional diagnostic x-ray tubes. Of the many types available, there are two major categories, tubes having stationary anodes and tubes having rotating anodes.

A stationary tube is composed of the previously mentioned basic components, with the anode in a fixed position; therefore the surface of the anode is constantly bombarded in the same spot by the electron stream. Although there are many uses for a stationary anode tube, the greater portion of this chapter will be devoted to the rotating anode concept.

ROTATING ANODE TUBES

Rotating anode tubes contain a tungsten disc, slightly mushroom in shape, which is rotated at approximately 3,300 rpm (Fig. 3-1). The newer triple-speed rotation anodes (10,000 rpm) allow rating increases (very high mA values with extremely short exposures) unheard of 10 years ago. A tungsten disc uses a ball bearing rotational system which must operate in a very high vacuum. Conventional lubricants will vaporize under these conditions and are unacceptable for rotating anode tubes. Metal film lubricants are deposited on the steel ball bearing to facilitate high-speed rotation. The stator of an x-ray tube utilizing a high-speed rotational system is provided with an electrical breaking apparatus. This reduces the high-speed rotation of the anode from 10,000 rpm to conventional speeds within several seconds after an exposure has been made. If the tubes were permitted to rotate for excessively long times at the 10,000 rpm range, severe strain would be placed on the bearings. Rotating anode tubes use an induction motor within the tube housing, although the bearings are within the tube insert.

STANDBY ILLUMINATION
OF THE FILAMENT

To avoid unnecessary tungsten evaporation of a filament, particularly at very high mA levels, radiographic tube filaments are put on a standby basis. When the unit is energized at any mA value, the filament is illuminated

Fig. 3-4. Rotating anode damage. Pitting, erosion, or cracking of an anode can occur if the manufacturer's rating charts are ignored. The anodes shown in this illustration have been damaged by instantaneous or chronic abuses. If an anode fails to rotate during an exposure, a "bullet-like" melt can result. See bottom disc. (Machlett Laboratories, Stamford, Conn.)

to a significantly decreased degreee than that required to produce the larger amounts of free electrons for high mA technics. When the actual exposure is to be made, a switch is depressed which energizes the stator of the radiographic tube, causing the anode to rotate. While the anode is being boosted to its proper revolutions per minute, the filament is raised from its standby level of illumination to the desired mA value.

Fluoroscopic procedures utilize low milliampere values, generally on the small focal spot of the under-table x-ray tube. When a radiographic spot film is desired, a cassette is brought into place by the fluoroscopist. In the time that it takes the cassette to travel from its park position to its expose position, the large focal spot of the x-ray tube is illuminated to the proper brilliance of a predetermined mA value. Simultaneously, the stator of the under-table tube boosts the anode to its proper speed, the exposure is made, and the cassette returns to the park position. Automatically, the large focal spot dims, and the small focal spot is boosted to the proper temperature to continue the fluoroscopic procedure.

Despite the filament protection afforded the technologist by the standby illumination of the filament, there are certain types of examination that occasionally require prolonged rotation of the anode and boosting of the filament. Mammographic procedures, which utilize high mA values (300mA) and long exposure times, can cause damage to the filament of an x-ray tube. It is sometimes difficult to use efficiently the high tube currents with the low kilovoltage values required for mammographic technics. The lower kilovoltage ranges (below 30 kVp) can be somewhat ineffective in wresting all the available electrons from a heated filament. There is a tendency to have the temperature of the filament increased to "boil off" more electrons. This can be a serious mistake if one is heating the filament beyond its designed limitations. Even when proper milliampere values are used without a noticeable drop in the milliampere reading, it is still possible to damage the filament of a mammographic tube by prolonged rotation of the anode, and therefore prolonged heating of the filament.

Occasionally, it can take an excessively long time to radiograph a difficult patient.

When this happens, the anode should not be continually rotated, for this places an unnecessary strain on the filament as well as on the anode, creating heat of an undetermined value within the tube. This extra generation of heat can damage a tube, particularly if a large number of exposures are to be made for a specific study, such as mammography. A similar problem occurs with serial vascular studies when the filament temperatures are boosted, anodes are brought up to proper speed, and then some type of delay occurs. It is always better to drop back to the standby illumination of the filament, and the de-energizing of the anode, than to risk tube damage.

Some mobile units feature a standby illumination of the filament, even though they do not have a rotating anode tube. When a timer value is decided upon and set at that value, the filament is automatically illuminated to full incandescence. It is important in operating a portable of this type that the timer station be selected just prior to the making of an exposure.

THE EFFECT OF TUBE VIBRATION ON FOCAL SPOT SIZE

A seemingly minor vibration in an x-ray tube during the making of an exposure can nullify the effect of the focal spot. Improperly balanced tubes, structurally unstable tube stands, or improper housing blowers can all add to vibration difficulties, with a corresponding degradation of the focal spot. Tube and crane stabilization must be checked periodically.

A last minute minor (frequently unnecessary) adjustment of a tube crane prior to an angiographic series can set up an almost imperceptible vibration of the tube, with resultant diminution of image sharpness.

Focal Spot Size Determination

Most conventional radiographic tubes have a "line-focus" effect (Fig. 3-1). Since the target is angled at approximately 15 to 17 degrees, the area that is bombarded by the

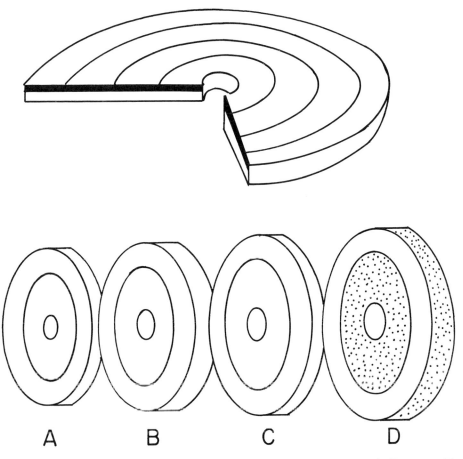

Fig. 3-5. Rotating anode composition and size. Most modern rotating anode discs are made of a "sandwich" of rhenium-tungsten alloy over a molybdenum body. A compound target usually has a greater heat storage capacity than does a pure tungsten target.

The size and thickness of a target influence the instantaneous as well as the heat storage potential of a tube. Targets A and B have identical instantaneous loading limitations because they are of the same diameter in size. Target B, because of increased thickness, would have a greater anode heat storage capability as compared with Target A. Target C has a greater instantaneous loading capability because of its increased diameter. Target B has a greater heat storage capability than does Target C. Target C, because of its larger size, can absorb more heat than Target A.

Target D represents a recent innovation in tube design. The entire anode body is made of graphite, which is considerably lighter than molybdenum. A ribbon of tungsten-rhenium is inserted into the periphery of the graphite disc by a special bonding method. The large diameter of the disc permits increased instantaneous loading, whereas the graphite body absorbs more heat and is able to dissipate heat at a faster rate.

---

electron stream is actually larger than the size of the projected focal spot.[11] The fact that the "line-focus" tube results in an almost square effective focal spot, even though the electron stream strikes an area that is rectangular in shape, helps both to dissipate the heat generated by the electron stream over the greater area of the target and to maintain a small effective focal spot. The electron stream is delivered along the lateral margins of the focal spot as 2 separate sources. This delivery of the electron stream as 2 almost completely separated energy beams protects the center of the focal spot from overheating (Fig. 3-1). The focal track is determined by the diameter of the disc being used. The larger the diameter of the disc, the more tungsten is available for bombardment by the electron stream, and therefore the longer is the focal track (Fig. 3-5). As the rotating anode moves, bringing fresh unbombarded tungsten into position, there is an increase in the instantaneous load capability of the tube.

There is general agreement that the smaller the focal spot, the better are the sharpness and definition of the radiographic image. But the smaller the focal spot, the more severe are the instantaneous exposure rating restrictions. Since there is a practical limit to focal spot size, a focal spot must be matched to the type of examination being considered. For example, if one is examining the chest of a patient who is unable to hold his breath, there is a definite disadvantage in the use of a smaller focal spot with lower tube loading capabilities, for if lower mA values are utilized, there must be a compensatory increase in the length of the exposure time. This is, of course, a disadvantage, and no benefit is to be gained if a highly defined beam is utilized with a moving object.

Focal Spot Size and the Tube

The more modern x-ray tubes consist of an anode composed of either solid tungsten or molybdenum coated with a rhenium-tungsten alloy (Fig. 3-5). A disc is mounted on a rather thin molybdenum stem that is joined to the rotor system. An exposure which generates an excessive amount of heat on the skin of the anode can produce a localized melted area of the focal track, or a slightly glazed area of surface melting around the periphery of the target. If this excessive generation of heat occurs when the target is stationary, a "bullet-like" melt or hole is produced within the target (Fig. 3-4). Many of the newer tubes have a compound target, a molybdenum base with a coating of rhenium-tungsten alloy. By adding rhenium to

the tungsten the anode is made more resistant to heat fissure. This rhenium-tungsten compound when bonded to molybdenum creates a heat storage capability considerably higher than that of the pure tungsten target. Although rhenium does not increase the actual instantaneous ratings of the tube, it may extend the useful life of the tube, for if there is less etching on the surface of a target, then fewer electrons are lost within the surface of the target. Since surface etching is reduced, radiation output should be maintained.

Interest in the type and size of the rotating anode disc has been considerable because of the desire to use higher tube loadings coupled with higher heat unit storage capabilities. Although the higher tube loadings and heat unit storage features are commendable accomplishments, the major benefit is the diminution in the size of the focal spot. Newer tubes permit not only higher instantaneous loading with shorter exposure times but also the use of smaller focal spots. Although the larger, faster rotating rhenium-tungsten discs have improved the focal spot capability of the tubes, the decrease in the conventional target angle, that is, the making of a "steep-angle" tube, has been the major manufacturing innovation resulting from the desire to diminish focal spot size (Fig. 3-3).

The major benefit realized with the newer "steep-angle" target has been the increase in load capability of a fractional-focus-spot tube. A "fractional-focus-spot" tube is defined as a tube having an effective focal spot of 0.3 mm or less. Chapter 4 is devoted to the use of the fractional-focus tube for direct roentgen enlargement technics, both conventional and angiographic. Although the decreasing of the target angle from the more conventional 15 or 17 degree angle to 10 degrees has increased the ratings of the radiographic tube, there has been a decrease in x-ray field coverage. Some of the 10 degree angle tubes are restricted to a field of approximately 14" × 14" at a 40 inch focus-film distance, due to the severe "heel" effect of the steep-angle tube. Because the steep-angle fractional-focus tube utilizes a rhenium-tungsten-molybdenum anode, rotating at 10,000 rpm, major improvements in the technical concepts of small vessel angiography have occurred. Selective and subselective catheterization procedures are now possible, for direct roentgen enlargement technics overcome some of the limitations of existing film and intensifying screen combinations.

Dual Focal Spot Tubes

All diagnostic x-ray tubes with rotating anodes are manufactured with 2 focal spots, generally of 2 different sizes, producing the so-called *small* and *large* effective focal spots. Each filament is mounted in its own focusing cup or metallic shield, which is used to help direct the path of the electron stream. Both filaments are mounted side by side in most radiographic tubes, and the decision to use one or the other filament is made at the radiographic control panel. Some radiographic tubes have 2 focal spots of the same size if the tube is intended for a specific use; for example, photofluorography of the chest. This is not unreasonable, for if one filament is damaged, the other filament can be put to use. Although most radiographic tubes are designed with filaments in a side-by-side position, some European tubes are manufactured with one filament above the other. Since such tubes utilize filaments one over the top of the other, the anode must have 2 separate focal tracks, one for each focal spot. A tube of this nature generally features a focal track using a steep-angle (approximately 10 degrees) and a more conventional target angle (15 to 17 degrees) on the periphery of the disc. A small focal spot is used with the steep-angle focal path and a large focal spot with the peripheral or larger focal track. Many of these tubes house a fractional-focus spot for use with the inner or steep-angle track (Fig. 3-6).

Typical Focal Spot Sizes

Since radiographic tubes are available with dual filaments, the technologist has the option of selecting effective focal spot combi-

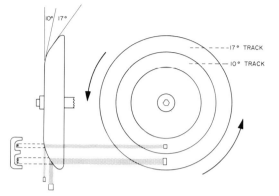

Fig. 3-6. Dual focal track concept. Most x-ray tubes contain 2 filaments, mounted side by side each of which is independently energized. Each individual electron stream strikes a common focal track.

A heavy-duty tube is available with 2 focal tracks as well as dual filaments. One filament is mounted on top of the other. The lower filament strikes the peripheral focal track (at a 15 to 17 degree angle) while the upper filament is used in conjunction with the inner track (at a 10 degree angle). There is a definite advantage to dual focal tracks, particularly if the "steep-angle" track is used for high-detail direct roentgen enlargement studies.

nations. Different focal spot combinations can be obtained from most tube manufacturers. Although we speak of 2 sizes of focal spots (small and large), it is probably better to divide focal spots into 4 groupings: (1) fractional focus, (2) small focus, (3) intermediate focus, and (4) large focus. "Fractional-focus-spot" tubes are available with 0.2 mm or 0.3 mm focal spots. Small focal spot sizes include 0.5, 0.6, and 1.0 mm selections. The intermediate size includes 1.2 mm and perhaps 1.5 mm, and the large focal spot size is generally considered to be 2 mm. While there are many other focal spot sizes available, the majority of radiographic tubes in use today have a 1.0 mm focal spot (small focus) and a 2.0 mm (large focus) combination. In general, a "small" focal spot refers to a 1.0 mm effective spot, and a "large" focal spot refers to a 2.0 mm focal spot. This 1.0

and 2.0 mm pairing will soon change with the newer tubes. Focal spots that are almost fractional in nature (0.5 and 0.6 mm) are in common use for cineradiographic technics, and can be used cautiously for some angiographic procedures.

Focal Spot Size versus Milliampere Values

It is interesting to note the milliampere values that are generally used with specific focal spot sizes. A comparison will be given of milliampere values applied to the 1.0 mm or small focal spot, as opposed to the 2.0 mm or large focal spot. The mA values selected are common to the more popular single-phase full-wave rectified 500 mA/125 kVp radiographic units.

1.0 MM FOCAL SPOT (SMALL)

	Occasionally
25 mA	150 mA
50 mA	
100 mA	

2.0 MM FOCAL SPOT (LARGE)

Occasionally	
100 mA	150 mA
	200 mA
	300 mA
	400 mA
	500 mA

The so-called small focal spot is restricted to one fifth the milliampere output of the x-ray unit (100 mA). This would mean that if the technologist were practicing high-detail or high-definition radiography using a small focal spot, he would be required to utilize relatively long, or at least moderate, exposure times. It is therefore fair to assume that the small focal spot must be restricted to relatively stationary segments of anatomy, such as the skull or the spine. Examinations where motion could be a problem—the heart, lungs, abdomen, or vascular opacification studies—necessitate the utilization of a larger focal spot so that high mA values can be used with shorter exposure times. While it is true that newer units and newer radiographic tubes permit the utilization of high mA values with small focal spots, these units are not in common use.

The most common technical change to avoid motion is the doubling of the milliampere factor and the halving of the time factor. For example, if an examination was being made using 100 mA at $\frac{1}{2}$ second (50 mAs), and it became necessary to shorten the radiographic exposure to overcome motion, the milliampere factor would be doubled to 200 mA, and the time factor lowered to $\frac{1}{4}$ of a second (50 mAs), with guaranteed duplication of radiographic density if the machine were properly calibrated. When this change is made, the effect of the small focal spot is lost. If sharp definition is required for a specific radiographic study, it is probably better to raise the kilovoltage value appropriately to overcome the halving of the radiographic exposure time. The raising of the kilovoltage value approximately 10 kVp in the moderate kilovoltage range approximately doubles radiographic density. An elevation of 10 kVp requires the use of a proper grid with extra attention to beam collimation.

TUBE RATINGS AND HEAT UNITS

Tube Rating Charts

Tube rating charts must be selected with the following considerations in mind: (1) the type of radiographic tube, (2) the focal spot of the specific tube in question, (3) the stator power (60-cycle or 180-cycle current), and (4) the type of rectification system used with the generator.

It must be assumed that when a rating chart is used, the values of the chart apply to reasonably cool tubes rather than to tubes that have just been subjected to a series of heavy exposures.

Basic Guidelines for the Use of Tube Ratings. These include the following:

1. Constant reference must be made to 3 charts:

A. The Tube Rating Chart (Fig. 3-7).

DYNAMAX "40", "46"

Fig. 3-7. Typical tube rating chart. Note the severe limitation placed on the tube by the use of the small focal spot instead of the large focal spot. Using an exposure value of 200 mA at 100 kVp and comparing the maximum permissible exposure lengths, the technologist finds that he is limited to $\frac{1}{10}$ of a second when using the small focal spot (1.0 mm). When employing the large focal spot (2.0 mm), he can use up to $3\frac{1}{2}$ seconds, an exposure value 35 times longer than that permitted on the small focal spot. As higher milliampere values are used at the same kilovoltage levels, tube ratings become even more restrictive. For example, although $3\frac{1}{2}$ seconds is permissible at 200 mA, only $\frac{3}{4}$ of a second is acceptable at 300 mA, and $\frac{3}{20}$ of a second at 400 mA; 500 mA cannot be used at the 100 kVp value. *Careful attention must be paid to the tube rating chart.* (Machlett Laboratories, Stamford, Conn.)

RATING CHARTS
60-CYCLE STATOR OPERATION

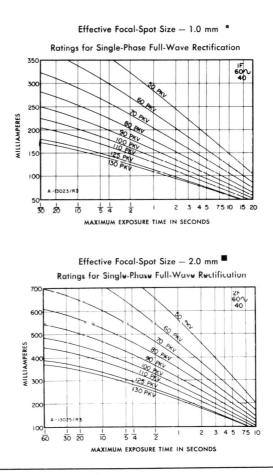

Effective Focal-Spot Size — 1.0 mm ▪
Ratings for Single-Phase Full-Wave Rectification

Effective Focal-Spot Size — 2.0 mm ▪
Ratings for Single-Phase Full-Wave Rectification

B. The Anode Thermal Characteristics Cooling Chart (Fig. 3-8).
C. The Housing Cooling Chart (Fig. 3-8).

Special charts are available for angiographic and cineradiographic studies.

2. The use of consecutive rapid exposures is quite risky, even though individual exposures are within the instantaneous ratings of the tube rating chart.

3. If higher milliampere values are used, an accurate heat unit determination is a *must* before a single radiographic exposure is initiated.

4. Housing cooling is rarely a problem during conventional diagnostic radiography.

5. Each radiographic tube has its own individual, totally different cooling chart. Modern day radiographic tubes with rotating anodes have anode heat unit tolerances varying from 72,000 to 400,000 heat units. It is not at all uncommon to find a heavy-duty angiographic tube (in excess of 200,000 heat units) installed in a neuro-angiographic suite side by side with a specialty head unit housing a smaller tube (72,000 to 110,000 heat unit capability).

6. Most portable or mobile units are se-

DYNAMAX "40",

COOLING CHARTS

ANODE THERMAL CHARACTERISTICS

RECTIFIED OPERATION

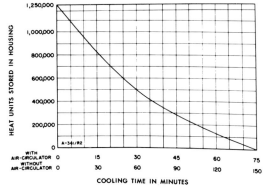

TIME IN MINUTES

HOUSING COOLING CHART

Fig. 3-8. Cooling charts. The anode thermal characteristic cooling chart must be constantly consulted, particularly when a prolonged study such as tomography is attempted. Because of concern for the instantaneous loading limitations of a tube, technologists tend to ignore the heat unit totals of a particular study.

Housing cooling is rarely a problem during conventional diagnostic procedures because of the generous heat storage capacities of modern day tube housings. (Machlett Laboratories, Stamford, Conn.)

verely limited in their ratings, and extreme caution must be maintained when these machines are employed.

7. Proficiency versus heat units spells caution. When technologists work together, general efficiency improves, and studies that seemingly took forever are quickly completed. The more competent the student or technologist becomes with a specific study or room, the greater is the risk to the radiographic tubes. The team concept of 2 tech-

nologists doing an extensive examination, such as a body section study, can be damaging to a radiographic tube.

8. Total exposures (total heat unit values) must be recorded. Each individual exposure must be tabulated as the study progresses, for if an error is made during an examination and the film is immediately discarded, the technologist must remember to add the heat units generated by an additional exposure to the heat unit total.

9. The use of an excessively high filament current setting for mammography will damage a tube. Filament current settings should be carefully set at the time of the installation of the tube, for if a high kilovoltage value is used with an excessively high filament setting, severe damage can result.[13]

Maximum Ratings for Individual Exposures

A tube should not be operated consistently at its maximum rating any more than any piece of equipment should be used at its maximum output. An automobile that is capable of 100 miles per hour should not be driven continually at 100 miles per hour. It follows that no technologist should continually use any x-ray tube at its maximum capability if he wants to avoid damage to the equipment. Roughening of the target can occur if there is continual use or abuse at the maximum rating of an x-ray tube (Fig. 3-4). A percentage of the electrons will be trapped within these roughened depressions; therefore a percentage of the radiation will be absorbed by the target. Erosion of radiographic quality almost parallels the erosion of the tungsten surface of the target. Radiation output can fall off as much as 50 per cent.

Heat Unit Formula: Single-Phase versus 3-Phase Rectification

A heat unit (H.U.) is defined as the energy produced by 1 kVp and 1 mA and 1 second in utilizing single-phase full-wave (2 pulse) radiographic equipment. Multiphase radiographic equipment (3-phase) requires 6 or 12 pulse current. With a 3-phase unit (6 pulse), multiply the formula by 1.35. With a 3-phase unit (12 pulse), multiply the formula by 1.41.

With an exposure technic of 80 kVp, 100 mA at 1 second, a comparison of the heat units on single- versus 3-phase equipment is shown:

Single-Phase Full-Wave (2 Pulse) Rectification

80 kVp × 100 mA × 1 sec = 8,000 heat units

3-Phase (6 Pulse) Rectification

8,000 H.U. × 1.35 = 10,800 heat units

3-Phase (12 Pulse) Rectification

8,000 H.U. × 1.41 = 11,280 heat units

It is never correct to assume that single-phase radiographic, cineradiographic, or angiographic charts can be converted to 3-phase ratings by the use of these formulas. The technologist must utilize the proper chart with the proper piece of equipment.

Generation of Heat Units

Whenever an x-ray exposure is made, heat units potentially damaging to the x-ray tube are generated. Since less than 1 per cent of the applied energy is converted into radiation, there is a tremendous amount of heat directly absorbed by the tungsten target of the tube. When a high-energy value is applied to the target in a single exposure, heat dissipation characteristics of the target area are such that this relatively high individual exposure may be used independently of the previous or subsequent exposures.[4] But a combination of such individual exposures results in collective heat units that can damage the radiographic tube. Even though safe single exposures can be made after the checking of the tube rating chart, heat continues to build up rapidly if consecutive exposures are made. The anode can melt, bend, or crack if subjected to this abuse. Many radiographic units are fitted with elaborate tube protective devices to safeguard the tube from overloading by individual exposures. *The overload system does not total heat units and is of no value when rapid sequence exposures are made.*

Housing Cooling. The heat that is generated by successive individual exposures to the tungsten target is transmitted rather quickly to the oil in the tube housing. Heat is then conducted away from the housing to the surrounding environment. Housing cooling is rarely a problem. The *Housing Cooling Chart* becomes important in a study such as a mass chest survey in which hundreds of patients are examined in a short time. The use of an air circulator to speed housing cooling adds nothing to the instantaneous exposure ratings of the tube.

Milliampere Seconds Conversions and Tube Ratings

It is easy to stress that the product of milliamperes times seconds results in milliampere seconds. Students continually are taught—and rightfully so—that 100 mA times 1 second results in 100 mAs, 200 mA times ½ second results in 100 mAs, and 500 mA times ⅕ of a second results in 100 mAs, and so on. Although the mathematics seems relatively simple, this formula cannot be applied practically to many x-ray tubes. The product of milliamperes times time is mathematically predictable, but the use of a large milliampere value with a comparatively short time is not always possible.

In using milliampere values above 200 mA, one must be cautious, for higher mA levels put serious limitations on tube ratings. In utilizing a Dynamax 40 tube (Machlett Laboratories, Inc.) at a 100 kilovoltage value, the maximum exposure value tolerated at 5 specific milliampere settings are shown. The large focal spot (2.0 mm) will be used for this comparison, and the anode is energized with a 60-cycle current.

able criticism from his employer if tube damage occurs. The technologist should constantly remind himself that even though a tube may be capable of accepting or tolerating exposures in the 200 mA, 100 kVp range, it is still imperative to total the heat units utilized in a specific examination. *Heat unit value must be totaled before beginning an examination.* Occasionally 6 or 8 exposures are made on a patient; each exposure may be within the individual exposure tolerance of the tube, but collectively these exposures are beyond the total heat unit capability of the tube. It is relatively easy to damage an x-ray tube by exceeding the heat unit capacity of the anode, particularly with studies requiring multiple projections. Some examples include tomographic procedures, serial films of a barium-filled stomach, etc. When 2 technologists work together, they frequently work faster than the anode can dissipate heat.

Kilovoltage versus Milliamperage: Their Effects on the Production of Heat Units

It is preferable, because of the heat units, to increase kilovoltage rather than milliampere seconds. For example, if one were

Milliampere Stations	Milliampere Seconds	Maximum Exposure Lengths
100 mA	1000 mAs	10 seconds
200 mA	600 mAs	3 seconds
300 mA	225 mAs	¾ of 1 second
400 mA	60 mAs	$\frac{3}{20}$ of 1 second
500 mA	Not possible (beyond rating of the tube)	

It is obvious why the technologist gets into serious difficulty if he insists on using higher milliampere stations. Increasing the kilovoltage level or, if need be, the length of exposure is better than raising the milliampere level in compensating for an extremely large patient or an inadequate technic chart.

It is a good general rule in using a single-phase full-wave rectified unit, to be concerned about exceeding a 200 mA or a 100 kVp value with an average x-ray tube. This suggestion does not apply to smaller x-ray tubes, particularly in older radiographic units. If a rating chart is not posted in the radiographic room, the technologist risks justifi-

using a technical value of 200 mA × 2 seconds × 85 kVp (single-phase full-wave rectification), a total of 34,000 heat units would be generated. If a 15 kVp increase were used with a 1 second exposure time (200 mA at 1 second and 100 kVp) as an alternate technic, there would be a heat unit total of 20,000 heat units, with a saving of 14,000 heat units. An increase from 85 kVp to 100 kVp would be based on the assumption that the radiographic tube could tolerate an individual exposure of this magnitude. With modern-day grids and collimators there is no justification in using low or moderate kilovoltage values to lessen scattered radiation.

Table 3-1. Some Typical Milliampere-Second and Kilovoltage Values

PROJECTION	FACTORS	HEAT UNITS	ADJUSTED FACTORS	ADJUSTED HEAT UNITS	HEAT UNIT SAVING
AP lumbar spine	100 mA-1 sec 80 kVp	8,000	100 mA-$\frac{1}{2}$ sec 90 kVp	4,500	3,500
Right oblique lumbar spine	100 mA-2 sec 80 kVp	16,000	100 mA-1 sec 90 kVp	9,000	7,000
Left oblique lumbar spine	100 mA-2 sec 80 kVp	16,000	100 mA-1 sec 90 kVp	9,000	7,000
AP tube angle 35 degree cephalad	100 mA-1 sec 85 kVp	8,500	100 mA-$\frac{1}{2}$ sec 95 kVp	4,750	3,750
Lateral lumbar spine	200 mA-2 sec 90 kVp	36,000	200 mA-1 sec 100 kVp	20,000	16,000
Lateral lumbo-sacral spot	200 mA-3 sec 95 kVp	57,000	200 mA-1$\frac{1}{2}$ sec 105 kVp	31,500	25,500
Total		141,500		78,750	62,750

Simple Technical Adjustment To Prevent the Generation of Excessive Heat Units

A simple technical adjustment to guard the radiographic tube from excessive heat input is the addition of 10 kVp to conventional technics with a halving of the mAs value. This general rule is certainly not 100 per cent accurate but is acceptable in many situations. Table 3-1 presents a list of some typical milliampere-second and kilovoltage values. These factors are not meant to be representative of any definite technic, but were formulated in the interests of mathematical simplicity.

Rapid Sequence Radiography

The most significant problem in the accumulation of heat units is posed by the efforts of technologists or student technologists working as a team, for a technical team can frequently work faster than an anode can dissipate heat. A good example is the usual rush to make follow-up radiographs after a fluoroscopic procedure. Stomach and duodenal films are usually phototimed at high kilovoltage values (125 kVp), with a high-ratio grid (12 to 1 or 16 to 1). If a 200 mA

station is used with a $\frac{1}{2}$ second exposure value on a relatively large patient, there is a heat unit total of 12,500 units per exposure with single-phase full-wave rectified equipment. Eight exposures of the duodenal bulb and antrum total 100,000 heat units in a matter of seconds. If the patient happens to be large, requiring a full second per exposure, a 25,000 heat unit per exposure value would be generated for 8 individual exposures (total, 200,000 heat units). Very few radiographic tubes can tolerate heat unit values of this magnitude.

It is common practice to divide a 10" × 12" or 11" × 14" film into 4 equal parts for tightly collimated views of the duodenal bulb and antrum. This severe limitation of the primary beam adds to the heat values produced, because if less scattered radiation is generated as a result of the severe beam collimation, there is decreased density on the film. The phototimer must ask for a longer exposure to maintain proper density.

In examining abnormally large patients it is better to take a single 8" × 10" or 10" × 12" film for each exposure. It is far better to waste film than to destroy the x-ray tube. As the primary beam is increased to a larger size, the exposure time can be reduced. Satisfactory radiographs with lower heat unit values are obtained. The primary beam should not be overly enlarged, for unnecessary radiation will be received by the patient. An added benefit of the larger primary beam size is the elimination of some involuntary motion of the stomach as a result of the shortened exposure time.

A similar situation occurs in attempting angiographic studies on larger patients. Since angiographic studies are not phototimed, the problem is not as severe, and one can easily determine the total heat unit values to be generated.

Single-Phase Full-Wave Rectification (Half-Wave Rectification from Damage to Valve Tube)

When it is necessary to operate a full-wave single-phase unit with a damaged valve tube, the lowest possible milliampere value should be used and then only as an emergency measure. A faulty full-wave rectification system should be repaired as soon as possible, for its operation puts an unnecessary strain on the filaments of the remaining valve tubes. The energies used at low mA values with a doubling of exposure lengths to compensate for half-wave rectification result in satisfactory film densities. When the faulty valve tube has been replaced, however, one can sometimes experience further valve tube failure in a relatively short time from the strain put on the rectification system by use of the full-wave rectified unit as a half-wave unit.

NEWER AND FUTURE CONCEPTS OF TUBES

Newer Rotating Anode Tube Concepts

Most rotating anode discs are 3 inches in diameter, although many newer tubes are manufactured with discs 4 inches in diameter. When a disc of larger diameter is used, considerably more tungsten is available for electron bombardment over a larger focal track without an increase in the size of the focal spot. As the disc increases in size, there is also a considerable increase in the heat storage capability of the disc. An increase in diameter or in the thickness of the anode can be used to increase the heat storage capability. There are discs 3 inches in diameter that are thicker than the conventional disc. This increase in the thickness of the disc increases the total heat storage capability of the disc, but does not increase the instantaneous loading capability of the disc. For example, the more conventional 4-inch disc of normal thickness can store approximately 200,000 heat units, whereas the smaller 3-inch but thicker anode can store 300,000 heat units. The larger 4-inch disc can tolerate a higher instantaneous exposure, but the smaller, thicker 3-inch disc can tolerate a greater number of individual exposures within the instantaneous loading capabilities of that tube (Fig. 3–5).

To increase tube output, the diameter as well as the thickness of the anode could be increased. There could also be some modification of the angle of the anode between the so-called 10 degree "steep-angle" and the conventional 15 to 17 degree target. An angle of approximately 12 degrees could be used with a conventional focus-film distance without a significant "heel" effect. The new targets could be made thicker, and could be rotated in excess of 10,000 rpm—for example, in the 13,000 rpm range. It is relatively easy to speculate about what should be done, but larger, heavier anodes create new rotational, ball bearing and vibration problems.

A new tube with a 4-inch diameter, 12 degree angle anode has been introduced by an American manufacturer. The entire anode is made of graphite, which is considerably lighter than molybdenum; therefore the target can be made larger. A ribbon of tungsten-rhenium is inserted into the graphite disc (Fig. 3–5). A special bonding method has been developed to ensure the adherence of the peripheral tungsten ribbon (focal track) to the graphite disc. This tungsten-rhenium ribbon graphite body has an improved thermal capability tolerating 400,000 heat units, with faster heat dissipation from the anode into the housing.

Future Tube Concepts

In a new tube design described by Dr. Michael Ter-Pogossian, a focal spot size is predicted of the diameter of 20 microns. This tube would be able to provide sufficient radiographic density for magnification technics. With this tube, electrons would be produced at the cathode by volatilizing its surface by means of a laser pulse. There would be a high x-ray output just before the target was destroyed. This apparatus would permit radiography of large body parts using a relatively low kilovoltage with exposure times as short as a few microseconds. A disposable tube of this nature is not improbable, and a tube of this type could be compared to a photographic flashbulb.[5]

DIAGNOSTIC X-RAY FILTERS

Any obstacle through which x-ray passes on its way from the focal spot to the object under study is called a filter.[15] The glass envelope of the tube, the oil insulation of the housing, and other housing material can act as "inherent" filtration.[14] The inherent filtration of most glass windows in x-ray tubes is equivalent to about one half millimeter of aluminum filtration.[7] Some of the newer x-ray tube assemblages have greater inherent filtration values than older x-ray tubes; therefore it is never safe to estimate the inherent filtration of a specific tube. If the technologist is concerned about the inherent filtration of an x-ray tube, he should discuss this subject with his equipment representative.

There is considerable interest in "additional" filtration or "added" filtration. "Total" filtration is equivalent to the sum of inherent filtration and added filtration.

Filters are added to increase the hardness of the x-ray beam. This is brought about by the interaction of the photons with the atoms of the filter material.[10] The most common filter material used with diagnostic x-ray equipment is aluminum. Federal, state or local regulations determine the amount of filtration necessary for specific types of radiographic equipment, although there is a general agreement that 2 to 3 mm of aluminum is adequate for diagnostic studies with the exception of the very lowest kilovoltage studies.

The major purpose of any type of filter is to attenuate the x-ray beam prior to its use. The absorbing material is inserted as close to the tube as possible.[12] If there are impurities in the filter, they are blurred out by the mounting of the attenuating material as close to the source of x-ray as possible. The primary beam emerges from the x-ray tube composed of photons ranging in energy from the selected level to extremely low values. Filters selectively remove the low energy photons that do not contribute to the penetrating effect of the x-ray beam. These softer x-rays would be completely absorbed by the patient and are of no value in the production of the radiographic image. Although the softer x-rays do not contribute to the finished radiograph, they do add to the radiation dosage received by the patient.

Filtration should be installed in the x-ray tube in a permanent or at least a semipermanent way. A filter should be difficult if not impossible, to remove. Many collimators have permanent built-in filters that cannot be removed. Some collimators have special translucent filters, which are removable, mounted in external tracks. These filters, spoken of as "aluminum equivalent" filters, are used with collimators that have an external set of lead shutters extending into the tube port for maximum control of off-focus radiation. These removable external "aluminum equivalent" filters frequently darken from repeated exposure to radiation or the light beam. They turn amber in color or, after prolonged usage, almost a dark brown, and it becomes very difficult to see the illuminated field size on the surface of the patient. It is easier to remove the filter from its track to visualize the light source on the surface of the patient than to shut off the room lights.

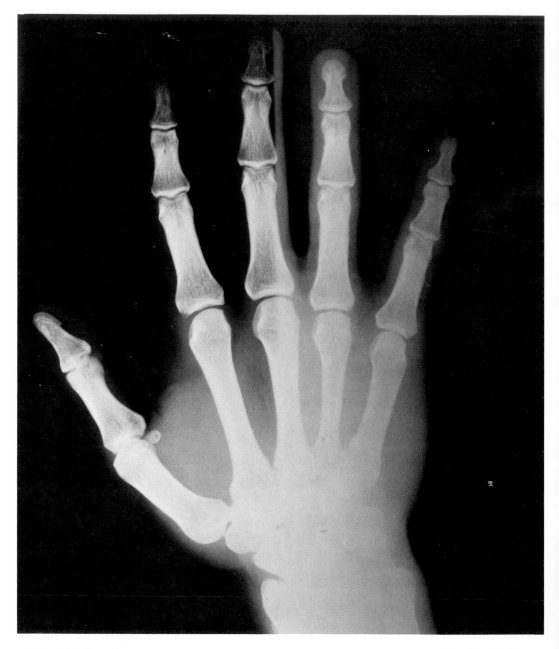

Fig. 3-9. The effect of added filtration on radiographic density. A conventional technic was used in exposing a film of a hand. Prior to the exposure the aluminum filter of the collimator was removed halfway from the mounting track. One half of the film was exposed to an unfiltered beam and is considerably darker than the properly filtered side of the radiograph.

One can accidentally make an exposure with the filter out, and subject the patient to soft, useless, and damaging radiation (Fig. 3-9).

Variable Thickness or Compensatory Filters

Compensatory filters overcome variations in patient anatomy, providing a more uniform density across the radiographic film. Some of the filtration materials used for variable thickness filters include barium-impregnated clay, opaque plastics, aluminum or copper step-wedges, liquid filtration systems, and so on. Some of these filters are made of step-like laminations of aluminum or copper, and produce a step-wedge effect of lines of demarcation throughout the entire radiograph. Wedge filters should be machined to produce a homogeneous density level on the finished film. Materials such as barium-impregnated clay, opaque plastics, or opaque liquid avoid a severe demarcation effect. The use of a wedge filter to expose effectively the foot from the toes to the tarsal area is an established technic (Fig. 3-10). The use of a bilateral wedge filter to produce an evenly exposed radiograph of the chest or the abdomen is not common practice. These filters, known as "trough" filters, can be of great help in the radiography of the chest or thoracic spine (Fig. 3-10).

With modern x-ray units specifically designed for chest radiography (a fixed focus-film distance, prealigned collimator, etc.) the wedge or trough filter prevents over exposure of the lateral aspects of the chest in the PA projection. With a specific-purpose unit it is even possible to incorporate a series of filters with a sort of "dial-a-filter" concept. A trough filter could be used for the conventional PA chest, a wedge filter for the lateral chest, and additional filters could be mounted on a disc-like device for attenuation of the beam on either side of the chest. For example, if the right lung field is completely radiopaque because of a tumorous mass or fluid, it is difficult to produce a chest radiograph with balanced density. The technologist is asked to examine a completely aerated radi-

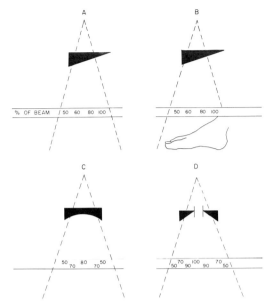

Fig. 3-10. Compensatory diagnostic x-ray filters. Filters of varying shapes can be used to balance radiographic film densities. A wedge filter (A) can be used in examining a foot in the AP position (B) to ensure an even density from the toes to the thicker tarsal area. Bilateral wedge or "trough" filters (C and D) help to compensate for density variations in the chest or thoracic spine.

olucent lung and a radiopaque hemithorax with the same x-ray beam. An opaque filter can effectively diminish the amount of radiation to the left lung so that a more balanced radiograph is achieved. Since the left lung is receiving less radiation, the phototiming device will continue to ask for more density. Exposure length will be increased, adequately penetrating the radiopaque right hemithorax. In a recent article Lynch recommends a wedge filter with the thickest portion of the wedge mounted on the superior portion of the collimator or cone for chest radiography, so that in the PA projection the thinner portion or the apex of the chest receives the least amount of radiation.[6] The wedge is reversed in the lateral projection so that the upper, more dense portion of the chest receives a greater amount of radiation than does the lower portion of the chest.

Placenta Localization. A useful technic with a compensatory opaque plastic filter for radiographic demonstration of the placenta is described by Cahoon.[1] Cahoon experimented with optimum kilovoltage technics and the anode "heel" effect in an effort to eliminate the use of the opaque plastic filter, but found that it was superior from the standpoint of overall rendition of detail in examination of the pregnant abdomen. McGann describes a semicircular double wedge of aluminum for placenta localization technics.[6]

Other Uses for Compensatory Filters. As the use of specialty equipment increases, compensatory filters should become more popular. The use of a compensatory filter is suggested in conjunction with urographic procedures. The lateral aspects of the abdomen would be more evenly exposed with the help of the trough filter since the abdomen in the AP projection is thicker in the center but thinner at its lateral aspects.

For the technologist interested in compensatory filters, a paper written by Thomas Funke in 1955 entitled "Compensating Filters in Medical Radiography" is *must* reading. Many of the newer concepts in radiographic technical literature were described by Funke over 15 years ago. These ideas are timeless and can be used with modern-day collimation devices.[3]

Funke used metals, plastics, and liquids in rubber bags as filtering agents, and classified filters as "underpart" or "portal" filters. An "underpart" filter used by Dr. Cesare-Gianturco consists of a 12 inch or 16 inch rubber compression bladder filled with a solution of contrast medium in water. The displacement of the opaque solution by the thicker parts of the area being examined permits most of the remnant radiation to reach the film. The absorption by the contrast material of much of the remnant radiation under the thinner portion of the part being examined produces a compensatory density effect. A rubber bladder filled with dilute contrast agent is helpful in examining the

knee in the lateral projection. A disadvantage of this method is that the part being examined receives the entire dosage. *The use of "portal" filters is recommended instead of "underpart" filters.*

The ultimate in a specialty filter was designed and described in 1934 by Arthur W. Fuchs for radiography of the entire body. A single exposure adequate to penetrate the lumbosacral spine, and still not overexpose the hands, ankles, and ribs, could be used because of the equalizing effect of the special filter. This filter is on display at the present time in the Museum of the American College of Radiology in Chicago, Illinois.[2]

REFERENCES

1. Cahoon, J. B., Jr.: Barium plastic filters in roentgen diagnosis of placenta praevia. X-ray Techn., *19*:185–188, January, 1948.
2. Fuchs, A. W.: Radiology of the entire body. Radiog. Clin. Photog., *10*:9–14, November 1934.
3. Funke, T.: Compensatory filters in medical-radiography. X-ray Techn., 27:12–18, July 1955.
4. Goodwin, P. N., Quimby, E. J., and Morgan, R. H.: Physical Foundations of Radiology. ed. 4. New York, Harper and Row, 1970.
5. Krabbenhoft, K. L.: Epilogue: the future of radiology. *In* Brecher, R. and E.: The Rays: A History of Radiology in the United States and Canada. p. 453. Baltimore, Williams & Wilkins, 1969.
6. Lynch, P.: A different approach to chest roentgenography: triad technique (high kilovoltage grid, wedge filter). Am. J. Roentgen., *93:965*, 1965.
7. McGann, M. J.: An aluminum filter for placenta visualization. X-ray Techn., *22*:304–305, March 1951.
8. Report of the Medical X-Ray Advisory Committee on Public Health Considerations in Medical Diagnostic Radiology (X-Rays). p. 8. U.S. Department of Health, Education, and Welfare; Washington, D.C., U.S. Government Printing Office, October 1967.
9. Selman, The Fundamentals of X-rays and Radium Physics. ed. 4, p. 50. Springfield, Ill., Charles C Thomas, 1970.
10. *Ibid.*, p. 155.

11. *Ibid.*, p. 190.
12. Stanton, L.: Basic Medical Radiation Physics. p. 100. New York, Meredith Corporation, Appleton, 1969.
13. *Ibid.*, p. 624.
14. Stanton, L., and Lightfoot, D. A.: The selection of optimum mammography technic. Radiology, *83*(3):442–454, September 1964.
15. Ter-Pogossian, M. M.: The Physical Aspects of Diagnostic Radiology. p. 156. New York, Harper and Row, 1967.
16. Van der Plaats, G. J.: Medical X-Ray Technique. ed. 3, p. 28. Eindhoven, Netherlands, Philips Technical Library, 1969.

4. Direct Roentgen Enlargement Technics

X-ray tubes with a "fractional-focus" spot (0.3 mm or less) have been available for direct roentgen magnification technics for nearly a quarter of a century.[4] These earlier tubes had severe rating restrictions, and their use was limited to magnification of thinner body parts. Although there was great interest in direct enlargement technics in the 1950's, it took nearly 10 years to accumulate an adequate list of articles describing the technical or physical principles of this procedure.[1,8,10,14,16,21,22,23] Some early investigators utilized the fractional-focus-spot tube for fluoroscopic or spot film radiography.[19]

ASPECTS OF ENLARGEMENT TECHNICS

Two Methods and Their Merits

Direct roentgen enlargement can be obtained in two ways, by (1) the tube-over-table method and (2) the fluoroscopic enlargement system.

Tube-over-table Method. For this a conventional x-ray unit is used, with the film holder placed 20 inches beneath the tabletop. The object being examined is placed midway between the tube and the film. A focal object distance (FOD) of 20 inches is used with an equal object-film distance (OFD) for a focus-film distance (FFD) of 40 inches. The resulting magnification using an equidistant focal object distance and object-film distance results in a 2× linear (4× area) enlargement. If the object is placed one third of the distance from the tube to the film, resulting in an object-film distance of two thirds of the focus-film distance—for example, a 13 inch FOD, and 26 or 27 inch OFD for a 40 inch FFD—a 3× linear (9× area) enlargement will be obtained (Fig. 4-1).

Fluoroscopic Enlargement System. A fractional-focus tube can be used for direct enlargement of fluoroscopic spot films (Figs.

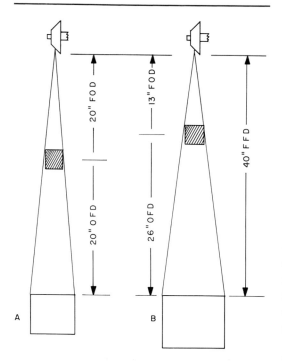

Fig. 4-1. Object-detector relationship for direct roentgen enlargement technics. If an object is placed midway between the tube and film (A), a 2× linear, 4× area enlargement will result. If the part under study is moved closer to the x-ray tube and farther away from the x-ray film (B), a greater degree of enlargement results. A 1:2 distance relationship will produce a 3× linear, 9× area enlargement.

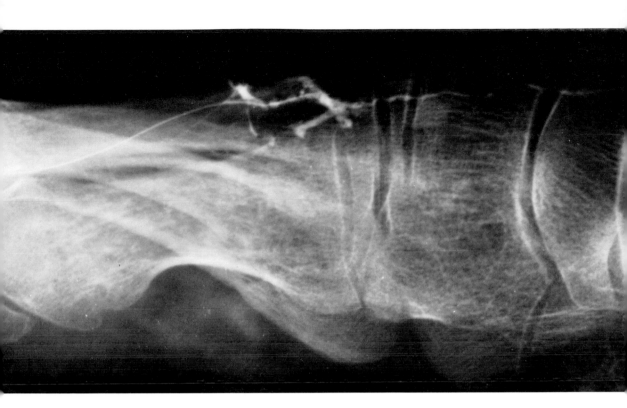

Fig. 4-2. Lymphangiography. A fluoroscopic spot film enlargement of the dorsum of the foot to guarantee the placement of the needle in a lymphatic vessel. Note the extremely fine catheter in position in a single lymphatic vessel.

4-2 and 4-3). An image intensifier must be used to facilitate positioning. When proper positioning has been ascertained, the fluoroscope is elevated to its maximum position, approximately 25 inches above the tabletop. Since the fluoroscopic tube is 18 to 20 inches below the tabletop (FOD), the object being evaluated is approximately midway between tube and film. The increase in the object-film distance from 20 inches to 25 inches over the conventional 2× enlargement technic results in a slightly greater degree of enlargement.[5,6,13]

There is a definite advantage to the fluoroscopic enlargement system, for the radiologist or the radiologic assistant is able to position a patient in a minimum amount of time and with great accuracy because of the fluoroscopic control. The fluoroscopic shutters permit exceptionally good collimation of the primary beam (Fig. 4-4). Many newer fluoroscopes combine under-table fractional-focus tubes with cineradiographic and 70 or 90 mm strip film cameras. With minor modifications to the control panel and spot film device, it is relatively easy to perform fluoroscopic spot enlargement technics.

Despite the deficiencies of a screen-film system, fast or very fast intensifying screens combined with fast-speed films can be utilized with the magnification technic. Minute radiographic details that are smaller than the resolving ability of conventional intensifying screens are magnified to the degree that they are perceptible. Spreading the radiographic image over a larger area with direct roentgen enlargement technics helps to overcome the deficiencies of the image detector system. The intensifying screens and films are recording larger bits of information with the enlargement technic than with the conventional technic. *As the degree of enlargement increases, the screen-film system becomes more efficient.* If an image of low contrast is excessively enlarged, it can consist only of radiographic penumbra. If the objects being enlarged are of high

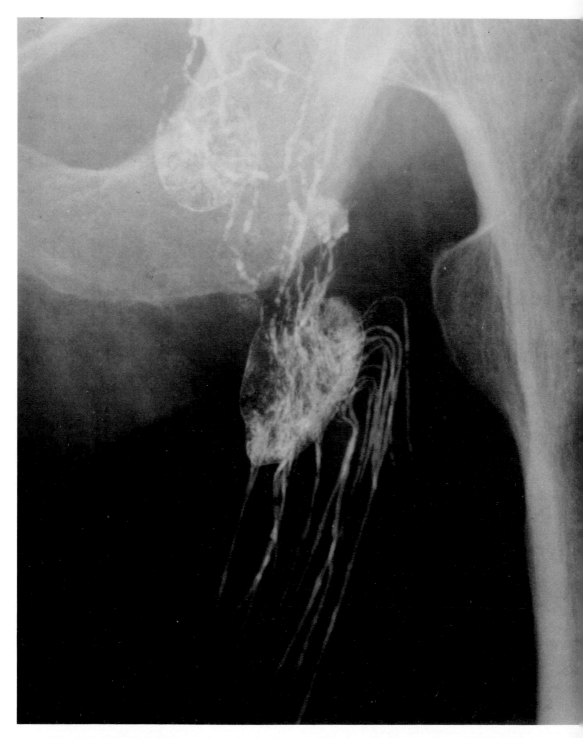

Fig. 4-3. Lymphangiogram. Lymph glands as well as lymphatics are demonstrated by means of a fractional-focus tube and a fluoroscopic spot-film device. The enlargement is 2× linear, 4× area. Factors used were 10 mA at 100 kVp for 2 seconds.

contrast, visibility of details can be enhanced; even though there is a significant increase in penumbra, there is an increase in the visibility of detail (Fig. 4-5).

Focal Spot Size and Penumbra

One should not attempt conventional direct enlargement roentgenography or direct enlargement angiography with a focal spot that is larger than the 0.3 mm fractional-focus tube. Users of radiographic tubes with small focal spots (0.5 or 0.6 mm) repeatedly attempt direct enlargement studies. The finished radiographs look quite satisfactory, and exhibit excellent radiographic contrast, for the scattered radiation clean-up mechanism of the air gap is not dependent on focal spot size. The radiographs appear to be diagnostic and most definitely are enlarged. *Small focal spots (0.5 mm, 0.6 mm) should not be considered adequate for direct enlargement roentgenography. Radiographic details that are smaller in size than the focal spot can "disappear" on an enlargement study.*

A minute radiographic detail seen on a conventional film can "disappear" on an enlargement film. For example, if a fracture were seen on a conventional film of the wrist, and if this fracture were smaller than the 0.3 mm focal spot of an enlargement tube, it is conceivable that the fracture would not be visible on a direct enlargement study. This fracture line might appear quite hazy and indistinct on a conventional radiograph, but nevertheless could be seen. At this point direct enlargement films may be taken, and used to support a misdiagnosis, for although there is a questionable fracture on the conventional film, the enlargement film leaves no doubt about the "normal" appearance of the wrist. This problem is quite serious when 0.5 and 0.6 mm focal spot tubes are erroneously used with enlargement technics. The use of a conventional focal spot (1.0 or 2.0 mm) results in the complete loss of image sharpness.

Scattered Radiation Reduction by "Air-gap" Method

Scattered radiation is generated in such a manner that it diffuses in all directions (360

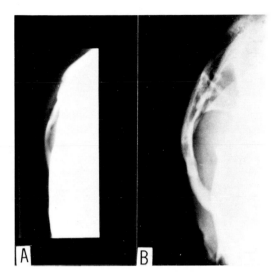

Fig. 4-4A and B. Direct roentgen enlargement fluoroscopic spot films. Conventional films in the SMV position of the zygomatic arch are difficult to obtain (A).

An excellent-quality spot film can be obtained by using an image intensifier combined with a fractional-focus tube as a positioning aid. The part under study is centered by means of the intensifier, the spot film tunnel is elevated to its maximum height, and an exposure is made. Proper rotation of the skull is quickly assured by fluoroscopic guidance. (B)

degrees). The intensity of scattered radiation is inversely proportional to the square of the distance. Although the weakening of the scattered radiation is controlled by the inverse square law, there is a negligible amount of absorption of the scattered radiation by air. Despite the fact that scattered radiation can expose a film in close contact with the object under study, a significant increase in object-film distance (the separating of the image detector from the object) has a surprising grid-type clean-up. If a longer object-film distance is used, an increasing percentage of scattered radiation in any given area misses the radiographic film (Fig. 4-6). McInnes claims that the clean-up with a 6 inch gap appears to be equivalent to that produced by a 6 to 1 ratio grid.[18] The use of a 6 inch air gap for enlargement radiography is not a technical advantage, for as the

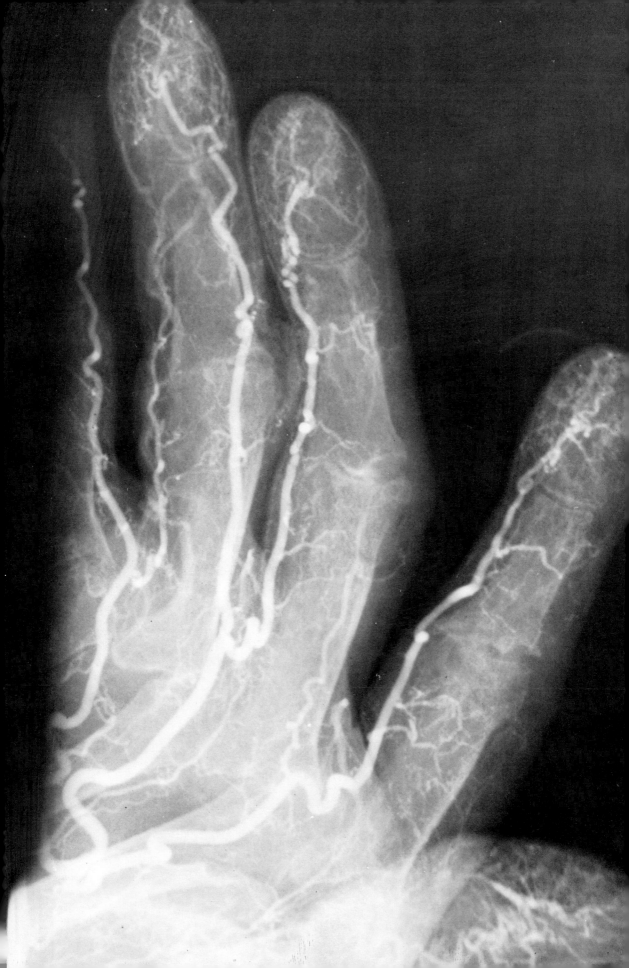

object-film distance is decreased, the focal object distance must also be decreased to maintain the proper enlargement relationship. If a shallow or shortened air gap is used, then a shortened focal object distance is required, with a corresponding increase in skin dose to the patient. The lessening of the focal object distance can be particularly disturbing in projections of the cranium. Lead shielding of the cornea should be used whenever possible if a shortened focal-object distance is used to examine the skull.

Kilovoltage Levels for Enlargement Technics

The use of an air gap permits moderately high or very high kilovoltage technics for enlargement studies. Many of the enlargement radiographs shown in this chapter were made with factors exceeding 100 kVp. All of the chest, abdominal, or skull enlargement angiograms were made with factors of 110 kVp or 120 kVp. High kilovoltage values are tolerated quite well by the fractional-focus tube. During an angiographic study there is a tendency to increase the milliampere values to produce the severe black-and-white effect required for a good contrast angiogram. It is not practical to exceed 125 mA in a moderate kilovoltage range when the newer fractional-focus-spot tubes are used, even though many of these tubes have a capability of 150 mA. Excellent subject contrast can be maintained with higher kilovoltage technics.

Clean-up of Scattered Radiation. Superb scattered radiation clean-up is possible with (1) severe limitation of the primary beam and (2) confinement of the primary beam to the part under study.

Severe Limitation of the Primary Beam. In the early days of enlargement radiography, there was a tendency to magnify areas of anatomic details out of proportion. Hands

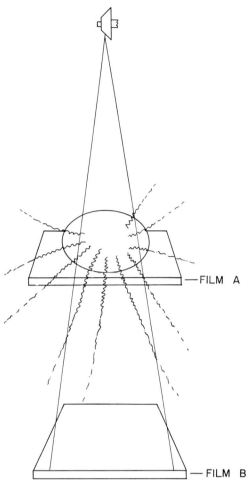

Fig. 4-6. "Air-gap" principle. The separating of the image detector, film, fluoroscope, etc., from the object has a surprising grid-type clean-up effect. Since film A is in direct contact with the part under study, most of the *scattered* radiation that has been generated will strike the film. When an air-gap is used (film B), because of its spatial effect, a high percentage of the scattered radiation completely misses the film. There is some absorption of the scattered radiation by air.

Fig. 4-5. High contrast enlargement angiographic study of the hand, 3 × linear, 9 × area. Note the lack of blood vessels feeding the 2nd phalangeal joint as a result of a disease process.

and feet were radiographed on 14″ × 17″ or 14″ × 36″ cassettes. The radiographs were visually dramatic, but of questionable diagnostic value. It is hard to believe that a radiologist could have difficulty in evaluating an entire hand and wrist, and would require an enlargement of an entire distal extremity. It

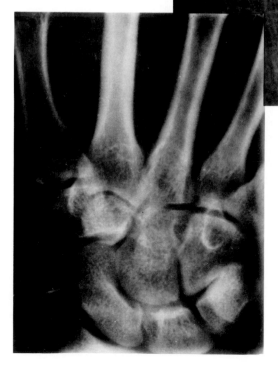

Fig. 4-7. A conventional view of the wrist (40 inch FFD) (*left*) **with a selective spot film enlargement of the carpal area,** 2 × linear, 4 × area (*above*). The enlargement radiograph was made with the aid of a fluoroscopic spot-film tunnel and an image intensifier. Multiple projections of an area of suspected pathology are possible with fluoroscopic guidance.

is more likely that a selective enlargement of a specific area of diagnostic interest, such as the navicular bone of the wrist, would be better radiologic practice (Fig. 4-7). It is better to take multiple enlargement views of the wrist to include the navicular bone on a single 8″ × 10″ or 10″ × 12″ radiograph than it is to enlarge the entire hand and wrist in a single plane to 14″ × 17″ size or larger.[13]

A similar selective technical policy can be applied to enlargement angiography. Many early studies were made of diseased areas. A better use for enlargement angiography would be the investigation of an area of questionable pathology noted on a conventional angiogram. The original conventional angiogram is evaluated, an enlargement study obtained, and a positive diagnosis is made, or there is strengthening of the negative or normal evaluation of the first study.

The *selective enlargement of a rather small segment of anatomy* will result in a large radiographic projection. A 4″ × 5″ (20 square inches) segment of a patient becomes 8″ × 10″ (80 square inches) when a 2× technic is used, and 12″ × 15″ (180 square inches) if a 3× technic is used. This means that if triple magnification technic is used with a 4″ × 5″ segment of a patient, a typical 11″ × 14″ cassette is not large enough to record the 12″ × 15″ projection (Fig. 4-8). If larger film or cassette changers were made available, it still would not be possible radiographically to evaluate large segments of the body, for as the primary beam size is enlarged, there is a significant increase in scattered radiation, and a grid becomes a necessity. It is impossible to overcome this increase in scattered radiation by conventional air-gap technics. A grid might be acceptable with

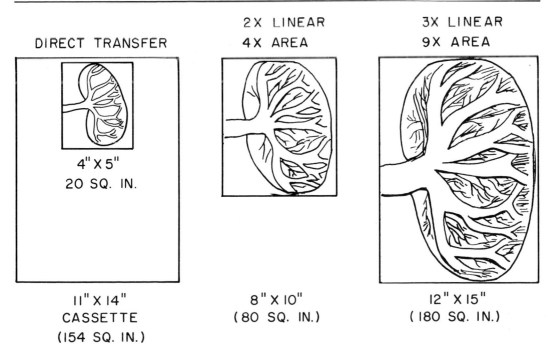

DIRECT TRANSFER

2X LINEAR
4X AREA

3X LINEAR
9X AREA

4″ X 5″
20 SQ. IN.

11″ X 14″
CASSETTE
(154 SQ. IN.)

8″ X 10″
(80 SQ. IN.)

12″ X 15″
(180 SQ. IN.)

Fig. 4-8. Frame size comparison between conventional and direct roentgen enlargement angiograms. Sizes shown are relative. A typical selective spot film angiogram is shown in which an 11″ x 14″ film was used (*left*). A 2× linear, 4× area enlargement study requires 80 square inches of film or 4 times the area required for a conventional angiogram. A 3× linear, 9× area enlargement requires 180 square inches of film, and is too large for a conventional 11″ x 14″ cassette.

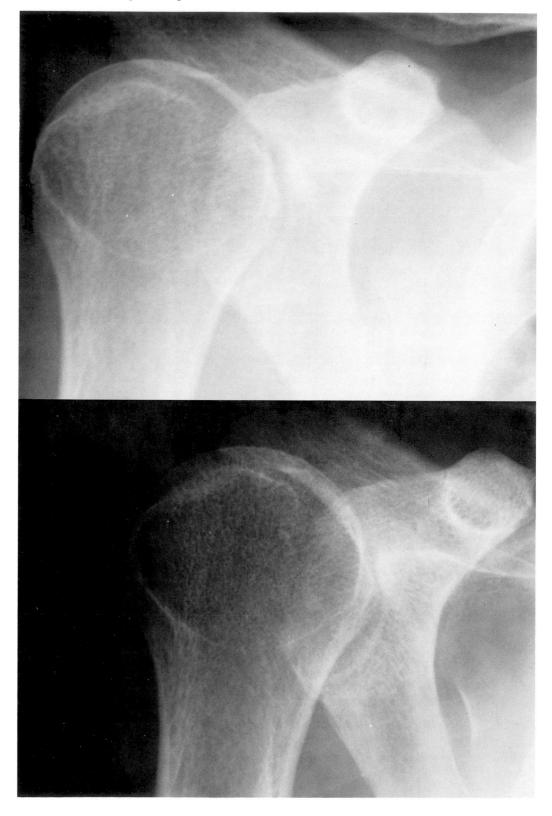

most routine enlargement studies but out of the question with angiographic procedures, in which motion is a constant problem.

Confinement of the Primary Beam to the Part under Study. If the primary beam is permitted to "leak" outside a body part, unattenuated radiation will strike the screen-film system. This can be particularly disturbing when the higher kilovoltage ranges are used to examine the skull, chest, or abdomen. Lead rubber shielding or sandbags can be used at the object site to attenuate the primary beam. For example, a study could be made of the left kidney with a 4″ × 5″ port for a selective enlargement renal angiograph, and the films could exhibit excellent contrast and detail. If with the same screen-film system, kilovoltage and milliampere factors, etc., a study of the same patient were attempted near the outer aspect of the left side of the abdomen to demonstrate the vessels of the spleen, the final films could be exceedingly gray. The increase in scattered radiation is due to the fact that the primary ray was permitted to leak at the lateral margin of the abdomen. Prior to an angiographic study of an area such as the spleen, a beam attenuator (sandbag, lead rubber, etc.) should be positioned at the object site to restrict the primary ray from striking the screen-film system. When one

examines thinner areas of the body, such as the extremities, this problem does not exist. Conventional enlargement films are shown to demonstrate the effect of a primary leak (Fig. 4-9).

Collimation Difficulties

To determine a primary beam size for enlargement angiography is simple since the film size of the serial changer is the limiting factor. Determining primary beam size is another matter when conventional enlargement studies are attempted. The tendency is to shape the variable rectangular shutters of the collimator to fit the part. Widening of the entry port results in an increase in scattered radiation. Severe restriction of the rectangular shutters results in excessively light radiographs. *Fixed collimator openings should be used for all enlargement technics.* A cassette of specific size should be decided upon for a specific area. The size of the cassette will determine the opening of the shutters of the collimator. This predetermined opening should be used for all studies that require a particular cassette size.

ENLARGEMENT ANGIOGRAPHIC TECHNICS

It is now possible to perform direct magnification angiography as a result of the development of fine-focus x-ray tubes that can tolerate rapid successive short exposures. Serial film or cassette changers are used in a manner similar to that of conventional angiographic studies, the limitations of the tube being the only restriction for programing of the serial enlargement procedure. Enhanced anatomic details are obtained by direct magnification. Extremely small vessels which at best are difficult to see on conventional angiograms are quickly identified. This technic is particularly helpful when small vessels that were not clearly defined on the conventional study must be evaluated (Fig. 4-10A and B).

A definite advantage is that the "blush" or the capillary phase of a conventional angio-

---

Fig. 4-9. The effect of a primary beam leak during enlargement radiography. Every effort must be made to restrict the primary beam to the part under study when high kilovoltage technic is used. A conventional enlargement (2× linear, 4× area) of the head of the humerus is shown to demonstrate the changes in film density that occur when the primary beam is not confined to the area being examined.

Both films of the shoulder were made with the same factors, 10 mA, 1 second, at 90 kVp. The primary beam was confined to the part in the upper frame. In the lower frame the tube was moved out of the field for less than 1 inch, resulting in a primary beam leak and a darker radiograph, produced by the increased scattered radiation.

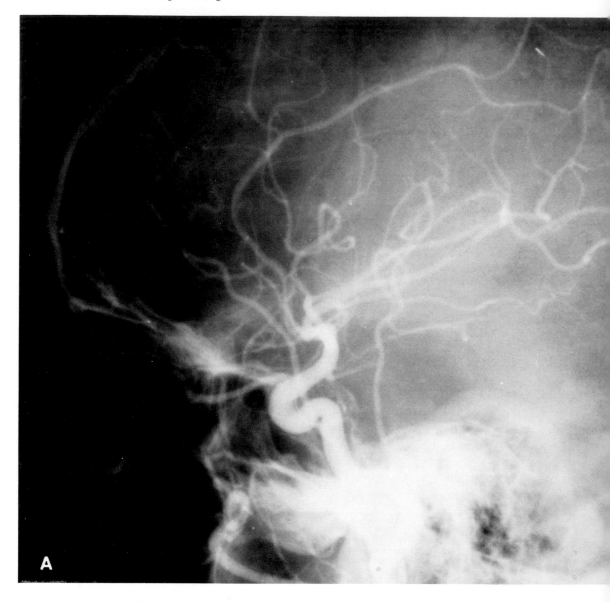

A

Fig. 4-10A. Conventional lateral cerebral angiogram, 2 films per second.

Fig. 4-10B. Direct roentgen enlargement angiogram, 3× linear, 9× area, 2 films per second. Note the wealth of increased vascular information.

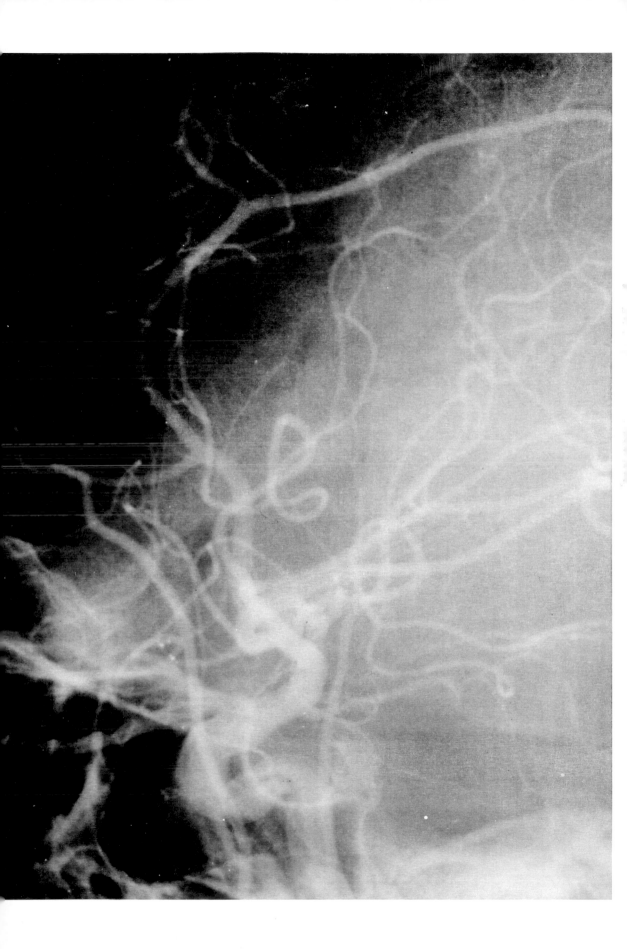

graphic procedure can usually be defined in the magnification films.[11] This "blush" is generally represented on a conventional radiographic film as a solid stain-like density. The enlargement technic can separate the "blush" into a myriad of radiographic details. Although it is true that some of these details are not in sharp focus, they are nevertheless there to see and to evaluate. A portion of this unsharpness is due to patient or vessel motion, which is accentuated (proportionately enlarged) by the direct enlargement technic. As newer radiographic tubes are fabricated with focal spots well below 0.3 mm, and newer target materials are developed to withstand higher instantaneous loads, further sharpening of enlargement images will be possible.

Recent Developments

In 1965, Greenspan combined angiography with the fractional-focus tube with excellent results.[9] Single-exposure angiograms were made of the heart, kidney, lungs, etc. In 1966, investigators at the Albert Einstein Medical Center in Philadelphia combined the enlargement angiographic technics with a

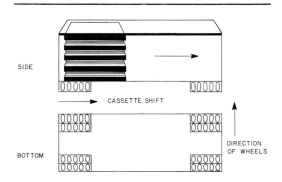

Fig. 4-11. Cassette changer stability. An appliance dolly consisting of 4 sections of 10 wheels each can be used to stabilize a 2 per second cassette changer for the "air-gap" technic. The changer is removed from its stand and placed on the dolly. The wheels should be so positioned that they will move in a direction opposite to the shifting of the cassettes. The grid must be removed for enlargement studies.

serial cassette changer.[6,12,15] There are technical problems with some serial film or cassette changers, such as poor screen-film contact, and rapid film sequencing helps to accentuate these difficulties. The possibility of considerable motion with excessive vibration within a film or cassette changer may add to image degradation. Screen-film contact and unit motion are common problems experienced with many film or cassette changers. The enlargement angiograms used in this chapter were obtained with a Sanchez-Perez Automatic Serialograph which had been removed from its stand and placed on an appliance dolly having 40 small nylon wheels. The dolly (4 sections of 10 wheels each) was installed with the wheels running opposite to the cassette shifting mechanism, producing excellent stabilization of the Sanchez-Perez unit (Fig. 4-11). The unit was used at its maximum cassette-per-second capability (2 per second). A 4-per-second cassette changer utilizing a vacuum system for the ultimate in screen-film contact is available for serial magnification studies.

Selection of a Table for Direct Roentgen Enlargement Angiography

One should consider the possibility of performing selective magnification when a special procedures table is purchased. If a large size serial changer is used, the top of the special procedures table must be capable of 20 inch elevation to produce a satisfactory air gap. If the table is not capable of being elevated for this procedure, a mechanism must be designed to lower the serial changer into the floor to achieve the 20 inch air gap for 2× linear (4× area) magnification. If 3× linear (9× area) magnification is contemplated, it might be necessary to use a thinner film or cassette changer in combination with a table that can be elevated. A 2-per-second cut film changer of a rather thin configuration is available with a special stand for enlargement angiography. This stand lowers the serial unit approximately 15 inches below the average tabletop. A 15 inch air gap can be satisfactory in some procedures, but if a 20 inch air gap is required, a table with elevation capability is a *must*.

If magnification cineradiography is contemplated, the input phosphor of the image intensifier must be able to be raised sufficiently to produce a 20 inch air gap.

Angiographic Research Possibilities

Some fractional-focus-spot tubes can be used to achieve 3× to 4× magnification without severe damage to image details. Studies achieving 5× linear (25× area) or 6× linear (36× area) are a possibility if one is willing to accept some diminution of image sharpness (4-12A, B, and C).

The use of a fractional-focus-spot tube with a body section device is possible if mechanical difficulties such as vibration can be eliminated from a unit of this nature. It would then be possible to achieve direct enlarged tomographic studies.[6]

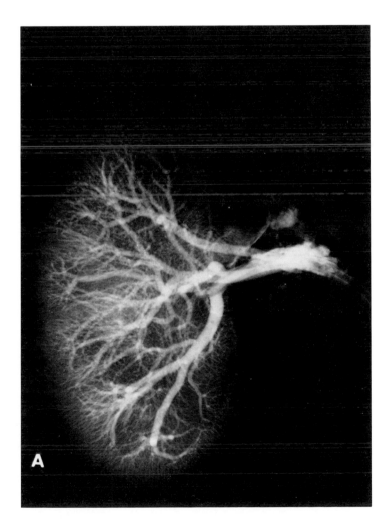

Fig. 4-12A. Conventional radiograph of an opacified specimen of a kidney. (*See following page for B and C.*)

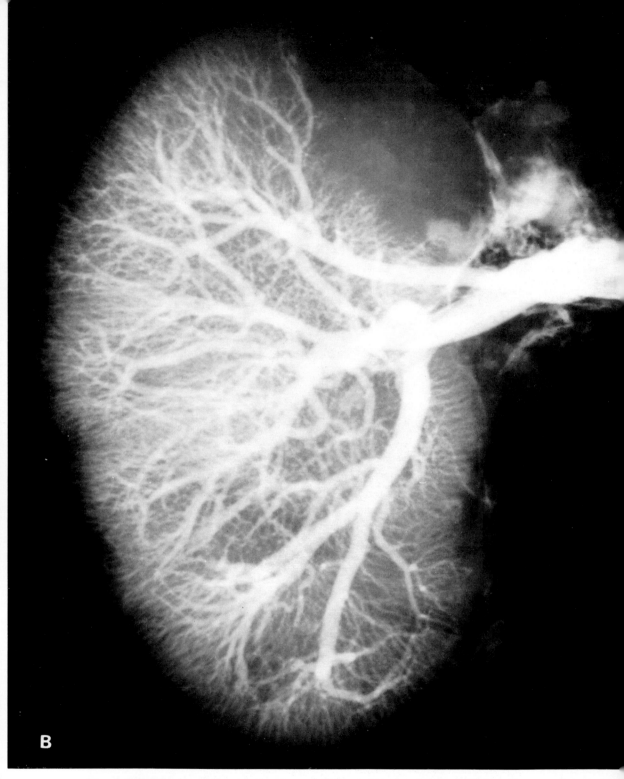

Fig. 4-12B. Direct roentgen enlargement of the same specimen as that in Figure 4-12A, 3×
linear, 9× area.

Fig. 4-12C. Direct roentgen enlargement of the original kidney specimen, 6× linear, 36×
area. Despite the unsharpness of this film, good subject contrast permits an enlargement of
this magnitude.

In practice, studies in the 3 to 4× linear, 9 to 16× area range are possible with modern
tubes.

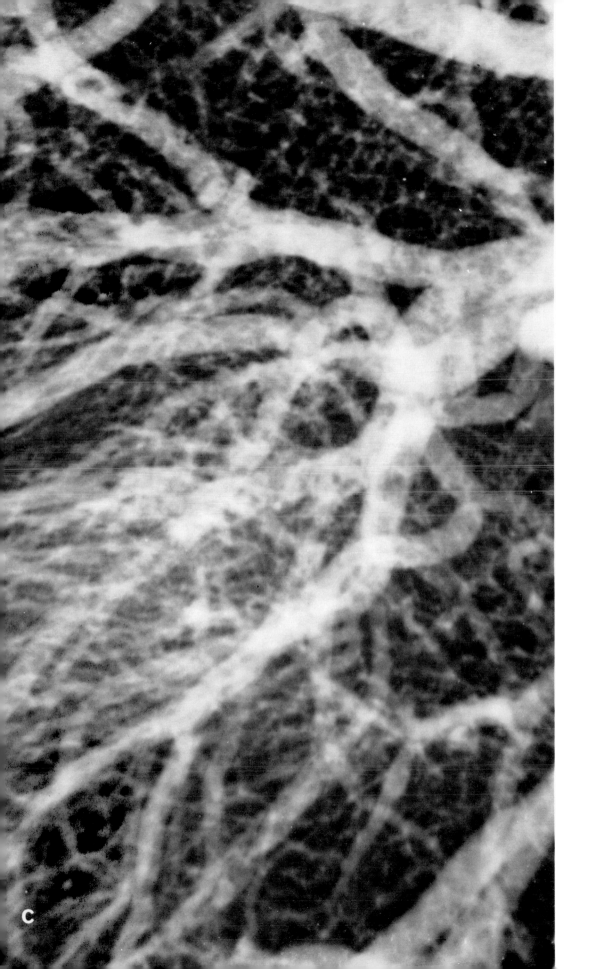

c

DIRECT ROENTGEN ENLARGEMENT
VERSUS
PHOTOGRAPHIC ENLARGEMENT

Attempts have been made to enlarge conventional films photographically, and direct roentgen enlargement radiographs have been carefully compared with the photographic enlargements. Greenspan compared the details of a direct enlarged coronary arteriogram with a photographically enlarged coronary arteriogram. Not only did the photographic enlargement enlarge the grain of the radiographic film, but the grid lines were also proportionately enlarged.[20]

Even if photographic enlargement technics were acceptable—*and they are not*—new technical problems arise. Very few radiology departments have the facilities to process, let alone enlarge, photographs. Valuable time would be lost with a photographic process.

Direct enlargement radiography, when combined with angiography, results in the immediate viewing of enlarged radiographs. If 90-second rapid processing facilities are available, more definitive projections or studies can be undertaken at once. *With photographic enlargement technics, the films are optically enlarged at a later time, and there is no possibility of additional films without subjecting the patient to an additional study at a later time.*

The use of *subtraction technics* in combination with conventional angiographic studies is increasing. This photographic procedure is quite time-consuming, and the time lost with the procedure must be compared with the time that would be lost with photographic enlargement technics. If one is an advocate of the subtraction method, the direct enlargement angiographic study can also be subtracted. Elaborate electronic systems are available for instantaneous subtraction technics.

GUIDES FOR NEW USERS OF THE
FRACTIONAL-FOCUS-SPOT TUBE

Each year brings new uses for the fractional-focus-spot tube—a tube that was thought of in the early '60's as primarily a research tool. Heavy-duty versions of these tubes are now commercially available from several manufacturers. Many technologists and physicians will soon be able to join the growing number of users of these tubes—and without guidance will probably duplicate all the technical mistakes made by the early users. An excellent bibliography on enlargement radiography has been published by Machlett Laboratories. Twenty-nine documented references cover technic, physics, and diagnosis.[7] It is recommended that the new user of a fractional-focus tube review this material prior to attempting enlargement radiography. Some excellent material that includes consideration of focal spot size has recently been made available.[1,2,3,9,11,17]

REFERENCES

1. Abel, M. S.: Advantages and limitations of the 0.3 mm focal spot tube for magnification and other technics. Radiology, 66:747–752, 1956.
2. Bookstein, J., and Steck, W.: Effective focal spot size. Radiology, 98:31–33, January 1971.
3. Bookstein, J., and Voegeli, E.: A critical analysis of magnification radiography. Radiology, 98:23–30, January 1971.
4. Burger, G. C. E., Combee, B., and van der Tunk, J. H.: X-Ray fluoroscopy with enlarged image. Philips Tech. Rev., 8:11, 321–352, November 1946.
5. Cullinan, J.: The role of the x-ray technician in the cardiopulmonary laboratory. X-ray Techn., 31:623–627 and 631, 1960.
6. Cullinan, J. E.: Fractional focus x-ray tubes; newer clinical and research applications. Radiol. Tech., 39:333–338, 1968.
7. Enlargement radiography, a bibliography. Cathode Press, 24(3):36, 1967.
8. Frantzell, A.: Effect of focal size, shape and structure on the roentgenographic representation of small calibre metal objects. Acta Radiol. (Stockholm): 35:265–276, April 1951.
9. Friedman, P. J., and Greenspan, R. H.; Observations on magnification radiography. Visualization of small blood vessels and determination of focal spot size. Radiology, 92:549–557, 1969.
10. Garnes, G.: Direct Exposure Enlargement Techniques Utilizing Fractional Focus X-Ray

Tubes. X-ray Techn., *23*:5, 323–328, 365, March 1952.

11. Hale, J., and Mishkin M. M.: Serial direct magnification cerebral angiography. Theoretical aspects. Amer. J. Roentgen., *107*:616–621, 1969.

12. Isard, H. J., Cullinan, J. E., Cope, C., and Leeds, N. E.: An Atlas of Serial Magnification Roentgenography. A Cathode Press Supplement. Stamford, Conn., The Machlett Laboratories, 1968.

13. Isard, H. J., Ostrum, B. J., and Cullinan, J. E.: Magnification roentgenography. Med. Radiogr. Photogr., *38*(3):92, 1962.

14. Jones, H. H., and Mahoney, G.: Some practical applications for the fractional focus tube. X-ray Techn., *27*:23–27, July 1955.

15. Leeds, N. E., Isard, H. J., Goldberg, H., and Cullinan, J. E.: Serial magnification cerebral angiography. Radiology, *90*:1171–1175, 1968.

16. Lofstrom, J., and Warren, C. R.: Magnification techniques in radiography. X-ray Techn., *26*:161–165, November 1954.

17. Mattson, O.: Focal spot variations with exposure data—important variations in daily routine. Acta Radiol. (Stockholm), *7*:161–169, March 1968.

18. McInnes, J.: The elimination of scattered radiation. Radiography, *XXXVI*(426):141–142, June 1970.

19. Philips Medicamundi: Skull Positioning. Medicamundi (Eindhoven, Holland), 1(1): 4.

20. Simon, A. L., and Greenspan, R. H.: Magnification coronary arteriography: Part I, Normal. Clin. Radiol., *16*:414–416, 1965.

21. Sparks, O. J.: Optimum projections with a 0.3 mm focus. Cathode Press, *15*:15–20, February 1958 (reprinted from X-Ray Techn.).

22. Van der Plaats, G. J., and Fountaine, J.: Applications of the Technique of Radiologic Enlargement to Chronic Joint Ailments. J. Radiol. Electr., *32*:249–255, 1951.

23. Wood, E. H.: Preliminary observations regarding the value of a very fine focus tube in radiologic diagnosis. Radiology, *61*: 382–390, September 1953.

5. Exposure Timing Devices

There are many types of timers that can initiate and terminate a radiographic exposure, including the standard spring-driven mechanical timer, which is the mainstay of many low output bedside units. The most available timing device, known as a synchronous timer, generally is driven by a synchronous motor and has a low range of approximately $\frac{1}{20}$ of 1 second. Electronic timing devices utilizing electronic tube systems, which have been available for some time, have timing increments as low as $\frac{1}{120}$ of a second. In recent years solid state devices have been replacing the electronic tube systems of the early electronic timers.

Regardless of the make of the timer used to initiate and terminate a conventional radiographic exposure, whether the unit be a simple spring-driven mechanical device or composed of complex solid state circuitry, a timer is set manually, the length of the exposure being predetermined by the user. The device is triggered manually and the exposure terminated at the end of the predetermined exposure value. Phototiming devices differ in that they determine the actual length of the radiographic exposure.

Understanding the basic concepts of phototiming devices is important, for automated timing, when it is used correctly, can improve film quality appreciably. Most of this chapter will be devoted to the principles of phototiming units that utilize complex mechanisms placed between the film and the source of radiation to terminate an exposure at a predetermined density.

As modern radiographic equipment becomes more complex, the need for accurate timers becomes increasingly apparent. Shorter exposure times are now an absolute *must* with new high milliampere output equipment in combination with fast intensifying screens and improved faster radiographic films.

TIMERS

Types of Timers

Spring-Driven Mechanical Timer. An extremely popular timing device is the spring-operated mechanical timer. This inexpensive timer is used with low-output radiographic equipment, such as portable or dental units, and is relatively accurate during long exposures. It is rarely manufactured to operate at $\frac{1}{10}$ of 1 second or below. It is quite adequate for routine bedside examinations, for low mA output equipment requires longer exposure times for adequate film blackening.

This type of timer can be a serious problem when used for bedside pediatric radiography. Although a low mA, moderate kVp bedside unit is more than adequate to penetrate the pediatric chest, the spring-driven exposure timing device is so insensitive that it is difficult to produce pediatric chest radiographs of good quality. Even when the timer consistently reproduces exposures in the $\frac{1}{10}$ of a second range, cardiac motion can plague the technologist, for $\frac{1}{10}$ of a second is equal to 12 individual impulses of current. Most pediatric radiographic units (300 mA to 500 mA) utilize $\frac{1}{120}$ of a second (1 impulse) or at most $\frac{1}{60}$ of a second (2 impulses) for radiography of the newborn chest.

Synchronous Timer. The motor-driven or

synchronous timer, the most common timer available, is far more efficient than the mechanical or spring-driven timer. It cannot be relied on for an exposure shorter than $\frac{1}{20}$ of a second.

Electronic or Impulse Timers. Timing devices which are basically electronic in nature are reliable up to $\frac{1}{120}$ of a second.[4] The latest electronic timer substitutes solid state devices for vacuum tubes. Some modern timers can control exposures that are a fraction of an impulse in length ($\frac{1}{120}$ sec, $\frac{1}{240}$ sec, $\frac{1}{360}$ sec).

Timer Increment Variations

Timer increment settings, while seemingly evenly spaced, are quite erratic. Slight adjustments from one timer increment to the next can produce severe density changes since a difference of 25 per cent of the initial exposure value is all that is required to produce a significantly perceptible change in radiographic density.[5] A wider selection of increment settings is needed, particularly in the ultrashort timer range.

Much x-ray equipment is severely—and regrettably—limited in regard to timer increments. Variations of as much as 100 per cent between timer increments exist (Fig. 5-1). A typical timer increment on a standard impulse timer varies from $\frac{1}{120}$ of a second (1 impulse) to $\frac{1}{60}$ of a second (2 impulses). The difference between $\frac{1}{120}$ of a second and $\frac{1}{60}$ of a second is 100 per cent.

Unfortunately, many technologists have the habit of increasing ever so slightly the kilovoltage or the timer value over the technic chart. For example, suppose a technologist were to radiograph a baby, utilizing $\frac{1}{60}$ of a second exposure time when the technic charts calls for $\frac{1}{120}$ of a second exposure time. It is as technically incorrect to radiograph the chest of a newborn infant utilizing this dou-

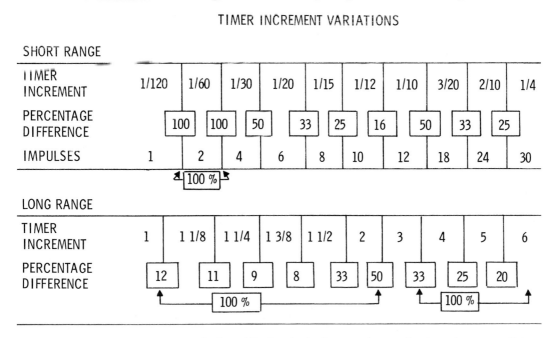

TIMER INCREMENT VARIATIONS

SHORT RANGE

TIMER INCREMENT	1/120	1/60	1/30	1/20	1/15	1/12	1/10	3/20	2/10	1/4
PERCENTAGE DIFFERENCE		100	100	50	33	25	16	50	33	25	
IMPULSES	1	2	4	6	8	10	12	18	24	30

100 %

LONG RANGE

TIMER INCREMENT	1	1 1/8	1 1/4	1 3/8	1 1/2	2	3	4	5	6
PERCENTAGE DIFFERENCE	12	11	9	8	33	50	33	25	20	

100 %   100 %

Fig. 5-1. Timer increment variation. Most modern radiographic timers are made with insufficient timer increment steps, particularly in the ultrashort exposure range. A variation in exposure value up to 100 per cent between stations is quite common ($\frac{1}{120}$ sec. to $\frac{1}{60}$ sec.). A single-step timer adjustment can result in a 100 per cent increase in exposure length, with a corresponding increase of 100 per cent in film density. In exposures of a longer range, 1 second or longer, timer increment values are not as critical.

ble-density technic as it would be to radiograph the lumbar spine of an adult in the lateral position using a 6-second exposure time when the technic chart calls for an exposure of 3 seconds. The difference between $\frac{1}{120}$ of a second (1 impulse) and $\frac{1}{60}$ of a second (2 impulses) has the same film blackening effect as the difference between a 3-second exposure and a 6-second exposure, an increase of 100 per cent in the radiographic value required. It would be unthinkable to increase an x-ray exposure from 3 seconds to 6 seconds, if the 3-second exposure were deemed adequate. And yet how often is an exposure on a newborn infant increased 1 increment on the timer, from $\frac{1}{120}$ of a second to $\frac{1}{60}$ of a second!

A Simple Method To Evaluate Timer Accuracy

Minor technical variations in the accuracy of a timing device are of little or no importance when longer exposure times are used. For example, if an exposure value of 1 second were used with conventional single-phase full-wave rectified radiographic equipment, 120 impulses of radiographic exposure would be the result. If the timing device were to vary plus or minus $\frac{1}{20}$ of 1 second (6 impulses), the 1 second exposure value could be as low as 114 impulses or as high as 126 impulses instead of the preselected 120 impulses. This small variation amounts to approximately 5 per cent (plus or minus) of the original exposure value. If $\frac{1}{10}$ of 1 second (12 impulses) were to be used, and the timing device varied (plus or minus) $\frac{1}{60}$ of a second (2 impulses), the actual length of exposure time could vary from 10 impulses to 14 impulses instead of the preselected 12 impulses. This approximately 16 per cent variation in exposure time could be technically annoying, but is still within the 25 per cent variation required for an obvious density change on a finished radiograph. A variation of plus or minus $\frac{1}{60}$ of a second (2 impulses) in using an exposure value of $\frac{1}{20}$ of a second (6 impulses) could be technically disastrous. This

error amounts to a 33 per cent variation in density from the original $\frac{1}{20}$ of a second (6 impulses) exposure value.

The technologist is able to check the accuracy of a timer by the utilization of a spinning top device. A circular disc of lead is provided with a small hole ($\frac{1}{16}$ of an inch in diameter). This disc is mounted on an axis so that it spins freely, is rotated, and an exposure is made. The individual pulsations of the tube current will be recorded on the film beneath the lead disc. One pulsation of current is required for each black dot visualized on the finished radiograph. If an exposure value of $\frac{1}{10}$ of a second (single-phase full-wave rectification) is utilized, 12 individual black dots should appear on the finished film; $\frac{1}{20}$ of a second should result in 6 individual black dots. If there is a failure in the rectification system resulting in single-phase nonrectified current, only the positive impulse of each cycle will be recorded as a black dot. One tenth of a second (nonrectified) is represented as 6 dots on the finished film; $\frac{1}{20}$ of a second (nonrectified) is represented as 3 dots on the finished film (Fig. 5-2).

It is important that the disc be rotated at a relatively moderate speed so that the black dots are not elongated and do not appear as light black or gray dashes. If the disc is rotated too slowly, the black dots will superimpose, making it difficult to determine the individual pulsations utilized per exposure. It is a simple matter of trial and error to select the proper speed. To test a timer, multiple exposures should be taken using the same exposure times. Variations in the accuracy of the timer can be easily determined if the dot values vary during multiple exposures at the same timer value.

Motor-driven spinning top devices are available for the evaluation of single-phase full-wave rectified radiographic equipment. A recent article by Cohen and Sterling demonstrates how to make a simple spinning top from materials costing less than a dollar. The device is quite accurate and can be made from a small lead disc, a simple toy gyroscope, and a large rubber eraser, which sup-

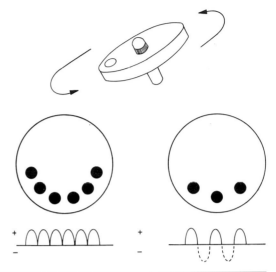

Fig. 5-2. Timer accuracy tests. The spinning lead top test is a reliable method for checking the accuracy of a timer of radiographic equipment utilizing single-phase current, whether full-wave rectified or half-wave rectified.

The spinning of the top during a conventional exposure results in a black dot for each single impulse of current. If an exposure of $\frac{1}{20}$ of a sec. were to be made, using full wave rectification, 6 black dots would appear on the finished radiograph.

If an exposure were made using the same time increment ($\frac{1}{20}$th of a sec.) on self-rectified or half-wave rectified equipment, the finished radiograph would record 3 dots. Spinning tops can be purchased or fabricated rather inexpensively.

ports the gyroscopic device for rotational purposes.[1]

Three-phase equipment requires sophisticated electronic equipment to evaluate timer accuracy.

Timer Recycle Time

Recycling time is determined by the time that elapses from the end of one exposure until the timer is ready to permit the next exposure.[6] This lapse of time is extremely important when the technologist is making a series of rapid exposures, such as a serial angiogram. Some timers have a recycling time of 6 to 12 exposures per second. When new radiographic equipment is installed in an angiographic room, the recycling time of the radiographic timer must be compatible with the type of film changer contemplated.

PHOTOTIMERS

Mechanisms

A complex device that reproduces radiographic exposures of consistent densities by automatically terminating an x-ray exposure is called a "phototimer." The radiographic exposure is initiated by the technologist, but the phototimer automatically controls the length of the exposure. The use of the phototimer can eliminate certain measurements

of the patient, but, contrary to popular belief, its use does not negate the skill of the technologist. The early users of phototimers felt that the technologist could be relieved of many responsibilities. On the contrary, more careful positioning habits are required, for radiographic positioning can be more critical with the phototimer mechanism than with a conventional radiographic timing device[2] (Fig. 5-3).

Phototimers are so efficiently accurate that repeat examinations result in duplicate errors. If a phototimer is in error, it will consistently repeat that error.

Structure and Action. A phototimer consists of a light-sensitive phototube which is optically connected to a fluorescent pickup screen. The photoelectric pickup of the phototimer incorporates a highly light-sensitive photocell or photomultiplier tube. A signal from the photomultiplier tube is amplified by an associated electronic circuit. This signal is made to terminate the radiographic exposure once a predetermined quantity of radiation has been delivered to the film[2] (Fig. 5-4). Some phototimer mechanisms use a plastic sheet behind the fluorescent pickup screen to collect the light given off by the pickup screen for evaluation by the photomultiplier cell.

Although there are many types of photo-

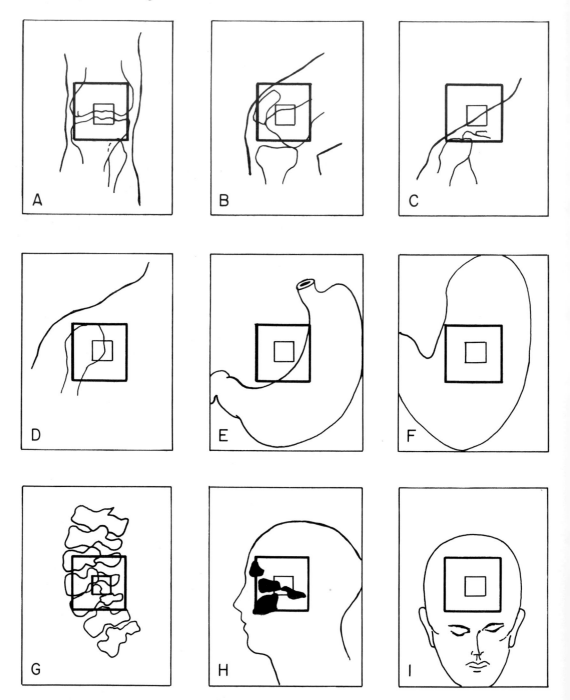

Fig. 5-3. Phototimer pick-up screen and positioning. A fluorescent pick-up screen or scanner is a vital component of the typical phototimer (see Fig. 5-5). Great emphasis has been placed on the use of an aperture of specific size for a specific part. While it is technically advantageous to select a specific aperture, either *large* (usually 4″ square) or *small* (usually 1½″ to 2″ square) for a specific body part, improper centering to the scanner will negate the function of the phototimer. The aperture must be properly centered to the part under study to achieve a predetermined film density. If the patient is not centered properly

timer mechanisms, the fluorescent screen photoelectric pickup and photomultiplier tube amplification device is most common. Another type of phototimer utilizes an ionization chamber placed between the film and patient.

The phototimer pickup device is usually situated behind the grid or Bucky. If it were to be placed in front of the grid or Bucky, serious sensing problems could occur, for the phototimer is asked to distinguish between remnant radiation and scattered radiation instead of remnant radiation. When a Bucky tray is used in front of a phototimer mechanism, a hole approximately 6 inches square must be available in the Bucky tray. This opening permits remnant radiation to pass unobstructed from the back of the cassette through the hole in the Bucky tray to the photosensitive pickup.

Cassettes with intensifying screens used for radiography in conjunction with phototiming have a thinner layer of lead foil behind the posterior intensifying screen than do conventional cassettes.

Scanning Discs and Later Systems. A multiple aperture scanning disc was used in the early versions of the Bucky phototimer. A single opening about 2 inches square (*small scan*), a group of 4 linear openings (*medium scan*) and a collection of 16 small circular openings (*large scan*) were used in conjunction with the phototimer to evaluate different segments of the body (Fig. 5-5). The small scan was used to evaluate relatively homogenous areas (skull, knee, etc.), the medium scan was intended to scan long, narrow anatomic segments (lateral dorsal and lumbar spines), and the large scan was required for areas of multiple density variations or areas of severe density differences (pelvis, abdomen, barium studies, etc.).

The scanning discs of the earlier models of the Bucky have given way to several systems. It is impossible to describe these systems without seemingly endorsing a particular device. Suffice it to say that the 4-slit or medium scan has been eliminated on all models. The small scan on many newer units can be 1 to 2 inches square, and the large scan is also a single opening, approximately 4 inches square. The same general positioning principles apply to the small scan and the large scan with modern phototimers (Fig. 5-3). Some units have eliminated the double scan system and use a single or universal scan.

A Review of the Workings of a Phototimer. The x-ray beam exits from the radiographic tube and penetrates the patient. Remnant radiation and scattered radiation pass through the table, striking the grid or Bucky. A considerable portion of the scattered radiation is eliminated because of the efficiency of the grid. Remnant radiation and residual scattered radiation continue to the cassette,

---

to the pick-up screen, overexposed or underexposed films will be the result. For example, in Figure A both apertures are centered correctly.

Figure B: The small scan is positioned correctly. The large scan is evaluating too much soft tissue.

Figure C: Both scans are situated poorly. Underexposed films would be the result.

Figure D: Both apertures would evaluate the bony structures of the shoulder correctly.

Figure E: Slightly underexposed films could be obtained, since both scans are under the soft-tissue structures of the abdomen rather than the dense barium-filled stomach.

Figure F: Since both apertures are scanning the barium-filled stomach, slightly overexposed films could occur.

Figure G: The patient should be moved slightly posterior to evaluate the lumbar spine with either scan in the lateral position.

Figure H: Both pick-up screens are scanning the air-filled sinus area. Underexposed skull films are likely.

Figure I: Both apertures are over the dense homogeneous skull vault. Either aperture should produce an adequate radiograph.

Although aperture size selection is important, proper centering to the pick-up screen is most critical.

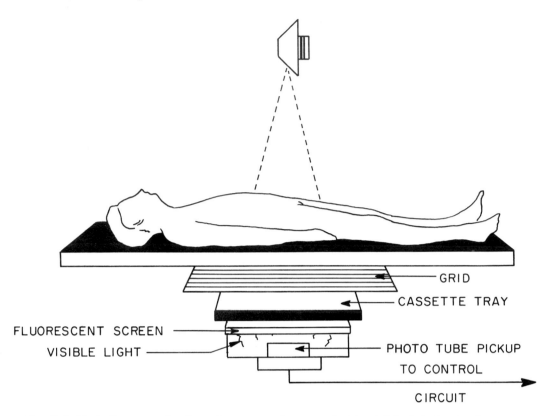

GRID

CASSETTE TRAY

FLUORESCENT SCREEN

VISIBLE LIGHT

PHOTO TUBE PICKUP

TO CONTROL

CIRCUIT

Fig. 5-4. Phototimer mechanism. *Remnant* radiation exits from the patient through the grid and exposes the film in the cassette tray. The x-ray beam continues through an opening in the cassette tray, striking a fluorescent screen in the phototimer mechanism. The x-ray is converted to visible light by the fluorescent screen. A highly light-sensitive photocell or photomultiplier sends a signal for amplification to an associated electronic circuit, which terminates the x-ray exposure at a predetermined value.

where the film is exposed. The remnant beam exits through the back of the cassette as a result of the thinness of the lead foil (0.0025 inches), passing through the opening in the Bucky tray, and strikes a fluorescent screen. The fluorescent screen gives off visible light. This light energy is picked up by the phototube and converted into electrical energy, which is transmitted to a control circuit. Predetermined settings within the circuit permit an electrical current to energize a relay that terminates the exposure.

This decision-making process by the phototimer mechanism can occur almost immediately, depending on the size of the object being studied. For instance, with a thin or relatively radiolucent segment of anatomy, the exposure can be terminated quite quickly. If a larger section of the body were to be examined, such as the lumbar spine in the lateral position, the exposure could be quite lengthy. These extreme differences in thickness and density lead to a discussion of two of the most widely misunderstood terms in phototimer radiography: "minimum reaction time" and "backup time."

Minimal Reaction Time

The reaction time or response time required by a phototimer to terminate the shortest possible exposure is known as the "minimal reaction time" of the phototimer.

The "minimal reaction time" is the shortest exposure time that can be obtained with a specific unit regardless of the intensity of the x-ray beam. This represents the shortest operating time of the phototimer system. *The minimum reaction time of a phototimer is independent of the quantity and quality of the x-ray beam.*

It is unfortunate that the minimal reaction time of many phototimer systems results in a longer exposure for a thin patient than might be necessary. For example, a very thin patient is to be radiographed by the use of a phototiming system with a typical grid chest radiographic technic (medium-speed intensifying screens, average-speed film, and a high ratio grid). A 200 mA station is selected at a 125 kVp value. If it is assumed that the minimum reaction time of this hypothetical phototimer is $\frac{1}{30}$ of a second—meaning that the phototimer in question cannot make a decision to terminate an exposure in less than $\frac{1}{30}$ of a second—then the patient receives at least the predetermined 200 mA, 125 kVp value at the minimum reaction time of the phototimer, $\frac{1}{30}$ of a second (6.6 mAs). But the minimum reaction time of this unit is too lengthy an exposure for the size of the patient.

Compensatory Adjustment Technics. Problems arising from a fixed minimal reaction time are sometimes very real, but there are compensatory technics. These will shortly be presented.

If the technical factors outlined above were retained in the instance of the hypothetical $\frac{1}{30}$ second minimal reaction phototimer, and, with all the original factors still holding, the patient were to step out of the primary ray as the exposure was made, the reaction time would remain at $\frac{1}{30}$ of a second (Fig. 5-6).

Frequently the technics are adjusted in less than desirable ways to compensate for difficulties with minimum reaction times. For example, if a 125 kVp high latitude, high penetration technic were being used, an adjustment might be made downward in the kilovoltage value (10–15 kVp) to try to decrease the total amount of radiation received

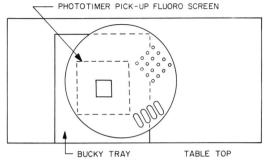

Fig. 5-5. Multiple aperture scanning disc. An early version of the Bucky phototimer required a multiple aperture scanning disc. Three apertures were used: *small*, a single opening; *medium*, 4 narrow openings; and *large*, 16 small holes over an area of 4 inches square. The medium scan was discontinued years ago, but the small and large (now a 4 inch square) openings are still in use. Some phototimers use a single or universal aperture.

by the phototimer so that, by increasing the length of the exposure, the phototimer could make an accurate decision beyond its minimum reaction time. But when the kilovoltage value is lowered, penetration of the part under study is decreased, along with radiographic latitude, and no longer is a high kilovoltage, high penetration chest film produced.

The minimum reaction time problem rarely occurs in the lateral projection of the chest as opposed to the PA projection. If a phototimer is consistently reproducing technics that are darker than desired in the PA chest projection, and if a high kVp, high penetration technic is desired, an adjustment in the milliamperage station from the original 200 mA to 100 mA or, in some cases, to 50 mA will help to overcome the minimum reaction time difficulty. Although the phototimer cannot react faster, the use of the lower milliampere station permits the phototimer to make its own decision concerning the length of the exposure over a longer period of time, which will at least exceed the minimum reaction time—in this case, $\frac{1}{30}$ of a second. If a 100 mA station is used with a

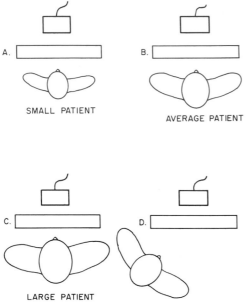

A.

SMALL PATIENT

B.

AVERAGE PATIENT

C.

LARGE PATIENT

D.

PATIENT OUT OF FIELD

Fig. 5-6. Minimal reaction time. The response time in making the shortest possible exposure with any phototimer is known as "minimal reaction time." Some units react in $\frac{1}{100}$ of a second, whereas others are as slow as $\frac{1}{30}$ of a second. Assume that the minimal reaction time of a specific phototimer is $\frac{1}{30}$ of a second, and the technologist decides to use this unit at 200 mA, 125 kVp, with *medium-speed* screens, average-speed film, and a high-ratio grid for chest radiography. It is possible that a patient of small stature (A) would require a shorter exposure time than the unit is capable of producing; an overexposed film would be the result. An average (B) or large patient (C) would require $\frac{1}{30}$ of a second or longer for proper film density. Good-quality films can be made on these patients. If a patient were to step out of the *primary* beam just as an exposure was made, the machine would still give off 200 mA, 125 kVp at its minimal reaction time, $\frac{1}{30}$ of a second.

$\frac{1}{30}$ of a second minimum reaction time, and the patient requires less than 100 mA at $\frac{1}{30}$ of a second at 125 kVp, an overexposed radiograph will still result. Occasionally, a patient is so thin or so severely emphysematous, or both, that a 50 mA station must be used with

a high kilovoltage technic. Lower mA values should be utilized with the well ventilated or emphysematous chest to give the phototimer the required time needed to make the proper decision.

If a Bucky is used with a phototimer mechanism, extremely short exposure times can produce grid striping patterns. It may be necessary to use a high-speed Bucky mechanism for extremely short exposure times. If for some reason the high-speed Bucky is objectionable, a fine line inorganic (aluminum) interspaced stationary grid should be considered to avoid grid striping impurities. In using a grid or Bucky technic, do not attempt to compensate for the minimum reaction time by increasing the focus-film distance beyond the range of the grid. As grid ratio increases, adherence to the recommended focus-film distance becomes more critical in both phototiming technics and in conventional technics.

Backup Timer

The "backup time" or the maximum exposure that a phototimer will permit is set within the ratings of the x-ray generator and, more importantly, within the limits of the radiographic tube in use. Standard tube rating charts are consulted to determine the backup time of any phototimer system.

Some classic technical errors that can destroy or at least seriously damage a radiographic tube if a backup timer were not incorporated into a phototiming system include:

1. The positioning of an x-ray tube in relation to a chest phototimer and the energizing of the Bucky table pickup. Primary radiation exits from the tube to the unactivated chest phototimer, while the phototimer within the Bucky mechanism waits in vain for a signal to terminate the exposure. The radiographic exposure will not terminate and will eventually destroy or at least seriously damage the radiographic tube.

2. The use of a phototimer setting in an attempt to take cross-table radiographs; for example, the lateral projection of the head

and neck of the femur, the transtable lateral decubitus film of the abdomen, chest, or gallbladder. Since a phototimer pickup unit is not available in the transtable position, serious damage to the x-ray tube can result.

3. The covering of a part of the patient by accident or by design with an opaque material, obscuring the anatomic area as well as the phototimer pickup. A lead apron used to protect a large segment of the body could produce this difficulty. A sandbag could be placed under an area, such as the femur, to support an injured leg, but it acts as an opaque barrier, completely or partially obscuring the phototimer pickup mechanism. A mistake of this nature, the obscuring of the photocell by an opaque substance without the use of a backup timer, results in a serious problem, whereas a mistake of this type during conventional radiography would result only in a poor diagnostic radiograph. Although the above-mentioned situations are without doubt bad radiographic practice, it is not uncommon to encounter mistakes of this type during student training. Probably all technologists have made at least one mistake of this type.

Phototiming Utilizing High kVp Technics

The use of high or moderately high kilovoltage values with an appropriate primary ray collimator and an appropriate ratio grid offers some advantage with phototiming as well as with conventional timing. The patient is exposed to less radiation, which can be of particular value in the examination of areas of the body where lower dosage is mandatory or repeat radiography is considered doubly dangerous. An example would be radiography of the pregnant abdomen; here higher kilovoltage values result in lower initial dosage and decrease the chance of repeat radiographs as a result of the increased latitude gained from high kVp radiography. An additional bonus is the utilization of lower tube currents; that is, lower mA values with corresponding smaller focal spots. Since mA values are held to moderately low levels, smaller focal spots may be

utilized. If high mA values are used with high kVp technics, exposure times are even shorter; but the combination of high kVp and high mA can be troublesome if a phototimer has a relatively long minimal reaction time. It is important to stress that even if the phototimer has an extremely short minimum reaction time ($\frac{1}{100}$ of a second), the technologist should be certain that a high mA, high kVp technic is within the ratings of the x-ray tube in use. *If high mA and high kVp values are utilized, the backup timer must be carefully adjusted to accord with the tube ratings.*

Phototiming Technic Priorities

Such examinations as chest or fluoroscopic spot films are particularly well suited to phototiming technics. The use of a carefully calibrated phototimer for chest radiography is almost a necessity. In a day and age when almost every large hospital routinely radiographs the chest of every admitted patient, it becomes increasingly difficult to justify the usage of a photofluorographic unit. Even the most modern photofluorographic unit requires a significant increase in exposure to the patient to expose adequately a 70 mm or 4″ × 5″ film. Modern film changers capable of transporting large sheets of radiographic film 14″ × 17″ in size can be operated with the efficiency of a 70 mm photofluorographic unit. These automated chest units can accomplish a great deal of work rather quickly. The incorporation of a phototimer system into a chest film changer not only enhances efficiency but is also a distinct advantage in attaining reproducible chest radiographs.

If a unit of this nature is not available, conventional phototiming chest units can be installed in existing radiographic rooms, and at specific times during the working day the units can be set up for routine but rapid chest radiography. It is frustrating to the busy technologist to have to break down or set up a room repeatedly because of a variation in work flow. Easy though it may be to criticize the technologist in charge of work assignments, actual scheduling of the work flow to suit departmental needs is often difficult. As hospitals enlarge and the distance between the typical patient room and the radiographic department increases, it becomes more and more difficult to schedule radiographic examinations efficiently; even the problem of an efficient escort service multiplies. In many large departments over 40 per cent of the patients are examined by conventional chest radiography. An automated chest system made as technically accurate as possible with a reliable phototiming system would mean that a patient could be radiographed and immediately sent back to his room or for another diagnostic test without concern for film quality. If chest films could be taken quickly with guaranteed density results, the escort attending the patient could be kept for the few minutes required for the examination; thus an additional telephone call for patient transportation and the invariable long wait for a patient escort would be eliminated. In short, the use of a phototimer for routine chest radiography has a dual purpose: (1) quality control and (2) departmental logistics.

Barium Studies. Phototiming is almost essential in examining the abdomen when barium has been used as a contrast agent. Although the installation of a phototimer mechanism in the Bucky table is desirable, the fluoroscopic device should be given first priority. It is the opinion of this author that phototiming for both the fluoroscope and the radiographic table is absolutely essential. One has only to look at a typical upper gastrointestinal or barium enema series to see the tremendous variation in densities that occur. During a routine barium enema examination one can see the descending colon (1 inch in diameter) and the cecum (5 inches in diameter) filled with barium, the cecum significantly more dense than the descending colon as a result of the increase in the amount of barium. A patient with a rather small-caliber colon can be examined with a high kilovoltage technic, resulting in a completely overexposed x-ray series. Another patient with a large-caliber colon can require extra-large amounts of barium for filling. A segment of the colon can be filled with gas and be completely radiolucent.

Conventional radiographic factors are woefully inadequate for the extra-dense or radiolucent colon. To say that the fluoroscopist should inform the technologist of any anatomic variant is easier said than done. Why should the fluoroscopist or the technologist be required to make technical decisions that automated equipment can make quicker and better?

Meeting Varied Needs. It is a common error in purchasing new radiographic-fluoroscopic equipment to install a complete phototiming system without regard to patient traffic flow. For example, every phototiming system must have a "brain" or decision maker within the panel. This phototiming master control is required regardless of the number of remote phototiming pick-up stations incorporated into the system. Multiple pickup cells can be installed in the room and be serviced by a single phototimer "brain" within the radiographic control panel.

Actually, pickup units should be installed with some priority in mind. For instance, a pickup cell can be incorporated within the fluoroscopic unit for use with a spot film device, a second pickup can be installed in the Bucky mechanism of the table, and a third unit can be installed in the chest Bucky; thus great technical flexibility can be guaranteed. Unfortunately, a room of this type will be used basically as a fluoroscopic unit, as opposed to a routine radiographic unit, since the completion of a department's fluoroscopic schedule by noon on any given day is considered a definite accomplishment.

The phototiming pickup cell within the fluoro-spot tunnel, and the phototiming pickup cell of the Bucky are used continuously, but the chest pickup unit remains idle, at least throughout the entire fluoroscopic schedule, for it is just too inconvenient to break down and set up a fluoroscopic room for conventional chest radiography. These modern high-powered radiographic units with complete phototiming systems are kept busy all morning, while the chest radiographic capability of the unit is virtually ignored.

Chest radiographs continue to be taken

with conventional radiographic equipment under less than ideal conditions. This situation would not exist if a department were completely phototimed; but this is not practical, and in many cases not necessary. To restress an important point, it would be wise

Position	Phototimer Station
PA	PA
PA	PA
PA	Lateral
PA	Lateral

to consider installation of a phototimer "brain" and single chest pickup cell in at least one radiographic room to handle the large daily volume of chest radiographs.

How To Use Known Variations In Phototimer Densities

Density levels of phototimer technics are generally determined by the manufacturer of the x-ray apparatus, although the user of the equipment is consulted at the time of installation. Let us use the term "density factor" to apply specifically to chest radiography and arbitrarily assign the number 1 as a density factor to the typical PA chest x-ray taken with a 72 inch focus-film distance with medium-speed intensifying screens and conventional radiographic film. If fast-speed intensifying screens were substituted for the PA projection utilizing the PA density station of the phototimer, an approximate doubling of radiographic density should occur, resulting in a density factor of 2. Many phototiming units are adjusted for the lateral projection of the chest to produce a density that is 2 to 3 times greater than the density required for a conventional PA projection (previously designated density factor 1). For the sake of simple mathematics, let us assume that one is working with a phototimer requiring for the lateral projection a tripling of density over that for the PA projection. This would mean that if a PA projection of the lung were to be taken, utilizing the lateral station of the phototimer with medium-speed intensifying screens, there would be a density factor of 3 on the finished PA radiograph. If a PA radiograph were to be taken, using the lateral

station of the phototimer and a cassette with fast-speed intensifying screens, the original density of the lateral would approximately double and produce a density factor of 6. The following is used to illustrate the density factor variation:

Intensifying Screen	Density Factor
Medium-Speed	1
Fast-speed	2
Medium-speed	3
Fast-speed	6

If this simple variation of known phototimer densities is used in combination with available intensifying screens, a technologist can continually reproduce 4 different densities of radiographic films in the PA chest position. Density levels 1 (original PA chest exposure) 2, 3, and 6 are produced with the distinct technical advantage that there is no change in the geometry of the radiographic image, and no change in the kilovoltage range (penetration) utilized for all 4 radiographs. With a fixed 125 kVp chest technic and a high-ratio stationary grid or Bucky, one is able to produce films of 4 different radiographic densities because of the phototimer density variables.

When one is asked to expose a Bucky chest radiograph, it is common practice to use AP dorsal spine technic. Several problems are generated by the substitution of a dorsal spine technic for a chest technic. Major geometric changes are encountered in using the conventional AP dorsal spine technic, for these films are generally taken with a 40 inch FFD, with the patient in the supine position. The conventional PA radiograph of the chest is usually taken with the patient erect and the anterior portion of the chest in intimate contact with the radiographic film. For the PA erect chest a 72 inch FFD is used for true teleoroentgenographic purposes, as opposed to the 40 inch FFD of the supine dorsal spine projection. If a questionable density is seen in the conventional chest radiograph in close proximity to or virtually superimposed on the mediastinum, it can be lost in the supine radiograph because of the enlargement of the mediastium, which is basically anterior to the film. Admittedly, there are times in the evaluation of certain chest conditions when this supine projection with its shortened FFD is more advantageous than the erect 72 inch PA chest film.

There is an additional technical disadvantage in exposing the chest with an AP dorsal spine technic when one is asked to produce a Bucky type radiograph of the lung fields. Dorsal spine technics frequently utilize exposures of 1 second or longer to achieve high mAs values. The look of a dorsal spine radiograph is significantly different from the look of a chest radiograph, for every attempt is made to accentuate the spine by a high contrast technic (high mAs with a moderate kilovoltage value). The longer exposure times needed to visualize the spine are an advantage in dorsal spine radiography, for normal breathing or pulsations of the heart blur out, or at least obliterate to some degree, lung markings adjacent to the dorsal spine. This helps to eliminate confusing vascular shadows. Radiography of the dorsal spine is designed primarily to emphasize bone visualization, whereas radiography of the chest is concerned with all of the structures within the thoracic cavity.

These density variations, 1, 2, 3, and 6, can be used for all types of chest radiographs. For example, if conventional oblique radiographs of the lung fields (between 5 and 15 degrees of rotation) are required for the localization of lung pathology, the PA station of the phototimer should be utilized with medium-speed intensifying screens. If oblique projections of the chest (45° RAO projection or the 45° LAO projection) are required for studies which utilize a very dense opaque medium, such as bronchography, a fast-speed intensifying screen could be substituted, and the PA station used for a double increase in density. When a barium swallow technic is used for a cardiac chamber evaluation, and small amounts of barium are superimposed on the dense mediastinum, the lateral station of the phototimer coupled with medium-speed intensifying screens is quite effective.

The use of 1, 2, 3, or 6 density technic can

be of some advantage in radiographing in the PA projection the chest of a patient with an enlarged heart or unusually well developed thorax. It is particularly helpful during bronchography in the PA projection to use the darker density settings when the trachea and the main stem bronchi are superimposed over the mediastinal structures.

A suggested technic for the PA bronchogram would require medium-speed intensifying screens with 1 exposure taken on a PA station for a conventional density chest radiograph to see the fine bronchial markings in the periphery of the lung, and a second PA chest exposure utilizing the lateral station for an overpenetrated, overexposed film with the same geometric proportions as the original PA radiograph. Remember that these films are teleoroentgenograms and can be compared with the initial survey radiographs.

The Use of Small Film Sizes in a Phototimed Chest Radiographic Unit

It is unfair to expect the chest phototimer mechanism to make a logical decision in using an 8″ × 10″ or 10″ × 12″ cassette for chest radiography, since the phototimer pickup cell is generally situated to evaluate a conventional 14″ × 17″ size field. By using a wood or plastic spacing device to elevate a small cassette into proper position in relation to the photoelectric pickup, a segment of the chest can be evaluated by the phototimer.

FLUOROSCOPIC TIMING DEVICE

Timing devices are required by law in certain states on fluoroscopic units. A predetermined time can be set on the fluoroscopic timer—for example, 3 minutes. The timer terminates the examination when the 3 minute period has been utilized by the fluoroscopist. These devices are invaluable training aids when physicians or radiologic assistants are learning the technics of fluoroscopy. They should be used with lengthy fluoroscopic procedures such as cardiac catheterization, where it is advisable to record prolonged exposure to radiation.[3]

REFERENCES

1. Cohen, S., and Sterling, S.: Personal communication.
2. Morel, J. M.: Reliability of phototiming in radiographic procedures. X-ray Techn., *36*(6): 402–405, May 1962.
3. Report of the Medical X-ray Advisory Committee on Public Health Considerations in Medical Diagnostic Radiology (X-Rays). U.S. Department of Health, Education, and Welfare; Washington, D.C., U.S. Government Printing Office, October 1967.
4. Ter-Pogossian, M. M.: The Physical Aspects of Diagnostic Radiology. p. 146. New York, Harper and Row, 1967.
5. Van der Plaats, G. J.: Medical X-ray Technique. ed. 3, pp. 354–355. Eindhoven, Netherlands. Philips Technical Library, 1969.
6. Vennes, C. H., and Watson, J. C.: Patient Care and Special Procedures in X-Ray Technology. p. 177. St. Louis, C. V. Mosby, 1964.

6. Technical Factors
—A General Review

The technical material in this chapter is philosophical in nature. Technical ranges rather than precise technical factors will be discussed. Whenever possible, technical generalizations rather than specific technical factors will be suggested.

For years the radiologic technologist has been accused of being a "button pusher." Certainly it is relatively easy to practice technology in a cookbook fashion. Technologists pass information freely from one to another in a recipe-like fashion. Much of this information is correct—and therefore useful; but misinformation is also perpetuated. The true "button pusher" develops a sort of know-how attitude about technology. Hard, cold technical facts are substituted for inventiveness. Technical innovation gives way to the boring repetition of adequate methods. The radiologic technologist who is truly technically oriented is not content to work with a know-how attitude but is more interested in a know-why attitude. To ply one's trade with a know-how attitude without the benefit of a know-why attitude is a simple matter; ideally, one should *know* both *why* and *how.*

Technologists and radiologists rarely agree on the relative merits of an x-ray technic. Disagreement is found among individuals as well as groups on the question "What is a diagnostic radiograph?" If a radiograph exhibits adequate penetration, and the organs are sharply outlined with good tissue and organ differentiation, it might be said to be a diagnostic radiograph.

A more practical definition would be "a radiograph which your radiologist considers to be diagnostic." The technologist can produce a radiograph of poor technical quality, and yet if a diagnosis can be made from this blurred, underexposed, or blackened film, he is congratulated for his technical skill. Diagnosis is in the eye of the beholder. Many a diagnostic radiograph can be retrieved from a discard-film barrel.

BASIC CONSIDERATIONS IN FORMULATING X-RAY TECHNICS

A major consideration in formulating an x-ray technic should be the control of radiation dosage to both the patient and the operator.[15] It is not the purpose of the author to dwell on radiation protection, but a review of some of the more important rules of radiation protection will be attempted.

How much exposure is too much to the patient? A very simple answer to a very direct question is that any exposure that a patient does not need is too much exposure. If a radiographic examination is ordered by a qualified physician, and is truly indicated, and if proper protective precautions are taken by the technologist in making the examination, the technologist ordinarily should not be concerned about radiation dosage.

Some simple rules to follow for all radiographic examinations include:

1. Do not be too quick to take additional views for your own convenience or the convenience of the x-ray department.

2. Females in the child-bearing age must always be questioned about their last menstrual period, for the possibility of pregnancy

must always be kept in mind. All patients with reproduction potential are entitled to proper gonadal shielding.

3. Whenever possible, radiographic technics using moderately high or high kilovoltage and low milliampere-second values should be used to lower radiation dosage. Fast radiographic films, in combination with fast intensifying screens, should be considered. Proper primary ray collimation is mandatory.

4. *A simple but accurate method for determining processor accuracy is necessary.* A report on "The Variability in the Automatic Processing of Medical X-ray Film"[20] stresses in its conclusions and recommendations that there is a considerable amount of variation in the base-fog densities of films developed in different automatic processors. It was also shown that these densities could vary considerably from day to day. The report stressed that because machines are considered to be automatic, it should not be assumed that they do not process films in an unvarying manner in a clinical environment. These authors recommend that in addition to performing routine maintenance tasks, periodic monitoring of the system performance be undertaken if the technologist wishes to provide radiographic images of consistent quality.[20] *It can never be assumed that a processor is functioning adequately unless a testing program is instituted and maintained* (Fig. 1-5).

5. A technic chart should be formulated with an awareness of the kilovoltage limitations of the grid.

6. Be certain that your radiographic unit is properly calibrated, institute a preventive maintenance program, if need be, and initiate simple calibration testing on your own, such as radiation output testing with a step wedge and timer accuracy determination using a spinning top[11] (Fig. 5-2).

7. Mechanical restraining devices should be used whenever possible. Differences in the use of restraining devices seem to be related mainly to the degree of interest and ingenuity of the physician or technician[15] (Fig. 2-28). Not enough attention is paid to the use of radiolucent positioning blocks, sandbags, compression bands, head clamps, and other positioning devices designed to help to eliminate motion.

8. Whenever possible, two views should be made at a right angle to each other in an attempt to achieve a 3-dimensional effect (Fig. 7-1). A listing of the number and types of projections for each examination should be made available to all members of a department.

9. Technical factors should be selected prior to the final positioning of the patient. Exceptions to this rule include the positioning of the patient in the upright or decubitus position for evaluation of the gallbladder with a horizontal beam technic. The patient should be positioned and several minutes allowed to elapse before the exposure is made, for small gallstones can be milling around within the opacified bile due to the vigorous movements required in positioning the patient. Waiting for a few minutes can be quite rewarding if a layer of calculi is visualized. If a last-minute adjustment of the patient is made, several minutes again should be permitted to pass before the horizontal beam radiograph is exposed.[2]

Another exception to the rule of calculation first, position second, is the making of upright or lateral decubitus radiographs to determine whether there is free air within the abdomen resulting from a ruptured hollow viscera. The patient should remain in the erect or lateral decubitus position for a few minutes to allow the free air to rise through the viscera of the abdomen to the diaphragm.

10. A good technical history is vital to the making of a quality radiograph. *Before exposing a single film, read the request, question the referring physician, secure as much information as possible, and look at the old films and reports.*[16] Often a technologist is required to make a "technical" diagnosis—for example, when he is asked to radiograph a long bone. Is the bone completely demineralized, requiring a decrease in kilovoltage, or is this structure extra-dense (due to a condition such as Paget's disease) and therefore difficult to penetrate? The

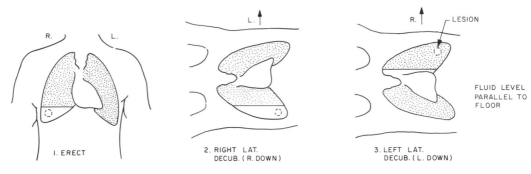

Fig. 6-1. Lateral decubitus radiography of the chest. A lesion in the base of the right lung can be obscured by fluid in the pleural space (1) when a patient is erect. If the patient is placed in the right lateral decubitus position (right side down), the lesion can still be obscured by the free fluid in the pleural space (2). When the patient is placed in the left lateral decubitus position (left side down), the fluid will often gravitate to the medial aspect of the pleural space, clearing the right lower segment of the chest (3) and so helping to evaluate the lesion.

It is absolutely essential that any film taken in an unusual position as described above or with an unusual tube angle be identified with appropriate lead markers. Overidentification of a radiographic film is impossible.

technologist must be able to decide when to increase or decrease conventional factors.

Exact adherence to the technic chart can result in technical chaos. For example, a patient can be referred for examination of the abdomen (measuring 22 cm), the technic chart is followed exactly, and the film can be hopelessly overexposed. Another patient (same size, age, and also with an abdomen measuring 22 cm) can be examined with the same technical factors, and the film can be woefully underexposed. The abdomen of the first patient (measuring 22 cm) could be distended due to an obstruction of the bowels, whereas the second patient could be an extremely muscular individual, or could have a dense fluid-filled abdomen, resulting in an underexposed film. The use of a photo-timer helps to eliminate these technical variations.

Table 6-1. Technical Grouping of 20 Body Areas*

Initial mAs Value	2× Initial mAs Value	4× Initial mAs Value
AP shoulder (bone detail)	Lateral skull	PA skull
AP cervical spine	AP or lateral femur	Towne skull

Lateral cervical spine
  A. 40″ Bucky or grid
  B. 72″ non-Bucky
	Tunnel knee (45 degree angle
	  Caudad)
	Ribs above diaphragm
		(plus 5 kV)
		SMV skull, base
		  (plus 10 kV)

Lateral sinus		AP abdomen
Lateral mandible		AP pelvis
Lateral oblique mandible (minus 5 kVp)		AP lumbar
AP lateral knee		AP dorsal
		Lateral dorsal
		Ribs below diaphragm

*This table is *not* to be considered accurate for routine radiography but was formulated to stress the similarities rather than the differences of technics.

A portable examination of the chest is a relatively simple matter. Again 2 patients can be evaluated with exactly the same centimeter measurements (AP diameter of the chest). Hyperventilation of the lungs or emphysema will result in a darkened radiograph. Pulmonary edema or advanced cardiac failure can result in an underexposed radiograph. Both patients can appear to be the same size, and can in fact be found to measure the same size, but actually require different technical factors. Do not minimize the taking of the technical history. A few minutes spent evaluating the situation can avoid a repeat examination.

11. Always identify a radiograph properly. It is impossible to overmark or overidentify a radiograph. One should develop a pattern for the proper marking of films. It is most unfortunate when a complete examination is marked correctly except for one unusual projection, such as a horizontal or tangential beam radiograph, which is inappropriately marked. An unusual view or tube angle can completely change the radiologist's approach to an interpretation (Figs. 6-1 and 6-2).

12. Maintain a fixed focus-film distance. A seemingly minor move, the lowering or raising of a tube (4 or 5 inches) can result in a major (plus or minus) density change on a finished radiograph, and play havoc with grid focus (Fig. 6-3). See in Chapter 7 the topic Variations in Focal Film Distance.

TECHNICAL FACTOR STANDARDIZATION

Too much emphasis has been put on minor variations in radiographic technic. An objective evaluation of a technic chart reveals that there is very little difference in technical values for many parts of the body (Fig. 6-4). Approximately 20 areas of the body can be placed in one of 3 mAs groups having only slight variations in kilovoltage. The first of the groups uses a fixed mAs value, the second group doubles the first mAs value, and the third group quadruples the first mAs value. For the sake of easy reading, let us assume

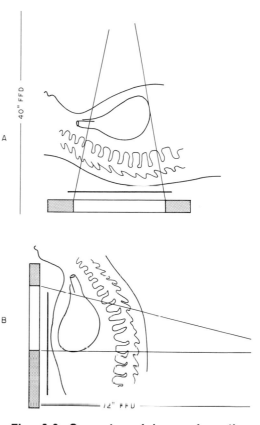

Fig. 6-2. Geometry of image formation. There are many ways of inadvertently altering a radiographic diagnosis by making a change in an established x-ray procedure. In Figure A the patient is in the AP supine position, and a 40 inch FFD is used. In Figure B the patient is in the PA erect position, and a 72 inch FFD is used. Note the severe enlargement of the cardiac shadow in Figure A. A widened cardiac shadow can mask a pulmonary hilar lesion. Technical changes of this nature may seem obvious, but they should be recorded on the x-ray request.

that column 1 requires a 10 mAs factor, column 2 a 20 mAs factor, and column 3 a 40 mAs factor. If it be further assumed that 80 kVp is the proper kilovoltage range to use with this mAs grouping in conjunction with specific screens, film, and grid (as determined by the user), then the 3 groups can be presented as in Table 6-1.

MILLIAMPERE SECONDS/DISTANCE RELATIONSHIPS

DISTANCE		MILLIAMPERE SECONDS		% INCREASE	
ORIGINAL	NEW	ORIGINAL	NEW	DISTANCE	MAS
20"	40"	10	40	100	400
20"	60"	10	90	300	900
20"	80"	10	160	400	1600
40"	60"	40	90	50	225
40"	80"	40	160	100	400
60"	80"	90	160	33	80

Fig. 6-3. Milliampere Seconds/Distance Relationships. A fixed focus-film distance (FFD) is a *must* for accurate radiologic technic. Changes in distance are not proportionately reflected in the changes required of milli- ampere second values. The lowering or the raising of a tube (4 or 5 inches) can be more damaging to radiographic density than tech- nologists usually care to admit.

This technical grouping can be further extended by variations (plus or minus) in kilovoltage.

There is no justification for a technic chart that has such fractional variations as these:

AP lumbar spine: 81 kVp, 100 mA at $9/10$ of a second.

AP pelvis: 83 kVp, 100 mA at $17/20$ of a second.

Ribs (below the diaphragm): 77 kVp, 100 mA at $8/10$ of a second.

PA skull: 79 kVp, 100 mA, at 1 second. These fractional variations are encouraged by the "one click" clique. Technologists and students make minor variations in technic, see their films, and feel certain that the obvious quality of the radiograph results from their adjustments to a "faulty" technic chart.

KILOVOLTAGE

Fixed versus Variable Kilovoltage

A fixed milliampere-second value coupled with variable kilovoltage values represents an early attempt to improve radiographic technic.[1] A simple rule to use with the variable kilovoltage technic states that an increase in the kilovoltage range by 10 kVp requires a 50 per cent reduction in the mAs value. This rule, which is inaccurate but adequate, is applicable in the range between 60 and 85 kVp.

Funke, in an article entitled "Pegged Kilovolt Technic," suggests a series of variations, successive steps in sequence, which deliver double the exposure to the film as compared with the preceding steps. Funke's pegged kilovoltage scale reads as follows: 44 kVp, 51 kVp, 57 kVp, 65 kVp, 75 kVp, 86 kVp, 100 kVp, and 120 kVp, with each kilovoltage step representing a doubling of exposure to the film. These observations are not educated guesses, but are grounded in scientific investigations. In a comparison of the kilovoltage differences in the 60 to 85 kVp range with the pegged kilovoltage scale (57 kVp, 65 kVp, 75 kVp, and 86 kVp), one observes in the pegged kilovolt scale a difference of 8 kVp between 57 and 65, and 10 kVp between 65 and 75, a difference that certainly is close enough to justify a 10 kVp variation for the halving of the mAs in this kilovoltage range. In the pegged kilovolt scale there is an 11 kVp difference between 75 and 86, again a difference that is quite adequate.

Funke also contends that the milliampere second value of any given technic should be doubled or reduced by one half for every 3 cm. of tissue variation up to 30 cm. Above 30 cm. he suggests 4 cm. variations as the rule.[8] After extensive testing McDaniel has reported a similar finding.[12]

Fuchs introduced the "optimum kilovoltage" technic, contending that for every body part there is an optimal or adequate kVp value for just the proper amount of penetration with the minimal amount of scattered radiation.[6,7]

Most American-made units are calibrated in 1 kilovoltage steps, whereas many foreign units are calibrated in 2, 3, 4, or 5 kVp in-

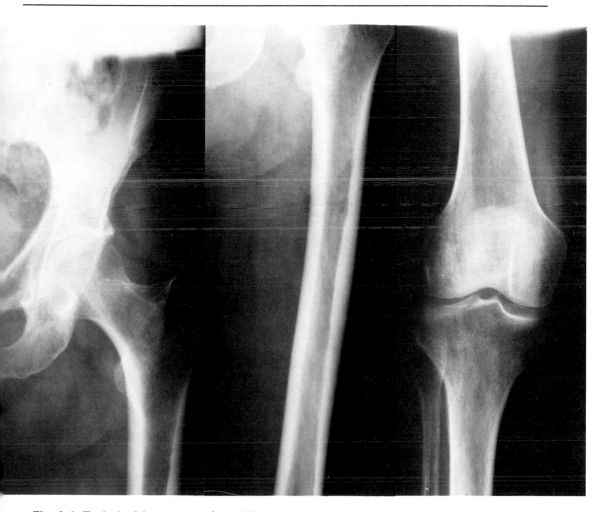

Fig. 6-4. Technical factor groupings. Minor variations in technic should be discouraged. These films of the hip, femur, and knee were all made at 70 kVp with a 12 to 1 Bucky; 25 mAs was used for the knee, 50 mAs for the femur, and 100 mAs for the hip. As body thickness or density doubled, the milliampere-second values were also doubled.

Note the shadow of the metallic needle added to the collimator shutters, and seen on the lower aspect of the hip radiograph as an unsharp opaque linear density.

crements. Some units vary in kilovoltage by steps of 2 kVp up to the 50 or 60 kVp range, where the kilovoltage values change to 3 kVp per step. After the 80 kVp range, increments of 4 or 5 kVp are utilized. The 5 kVp increment approach virtually eliminates the "one click" technologist. *An increase of 1 kVp, particularly in the medium to high kilovoltage range, is of no technical consequence.* An increase of 1 timer increment can be disastrous.

High Kilovoltage

The increased penetration resulting from higher kilovoltage values is technically advantageous. Although high kilovoltage studies exhibit a reduction in contrast, the film will contain details over a wide variety of density ranges or part thickness. A major advantage is the shortening of exposure lengths.[10]

Technics that require high contrast, such as intravenous urography, cholecystography, angiocardiography, or cerebral angiography, do not lend themselves well to high kilovoltage. But if motion is a problem in these studies, it is necessary on occasion to use moderately high kilovoltage values. A vascular study lacking something in contrast is obviously better than an extremely black and white series exhibiting motion.

Kilovoltage and Contrast

As the kilovoltage range increases, short-scale radiographic contrast suffers. Radiographs exhibiting severe areas of black and white are a complete waste of film. A blank area means simply that no exposure has reached the film; therefore no development has occurred, and unexposed silver crystals were cleared by the fixing process. In general, a long-scale contrast technic results in radiographs of more consistent diagnostic quality. In such radiographs the penetration of all areas is more uniform, and therefore the number of density levels for evaluation is considerably increased.

150 kVp Technic

The final value of the 150 kVp technic is yet to be determined. With a very high kilo-voltage technic, exposure values in the range of $1/120$ of a second are possible.[4,14] An additional bonus with the utilization of the 150 kVp technic is the significant reduction in heat units for any given safe exposure[13] (Fig. 6-5).

CONTRAST AGENTS

The Use of Radiopaque or Radiolucent Contrast Agents for Maximum Tissue Differentiation

Some anatomic areas are extremely dense, attenuating a considerable amount of the primary beam (Fig. 6-6). When an AP projection of a lumbar spine is processed, there is excellent differentiation in density between the dense bony vertebrae and the significantly less dense surrounding tissues. Good subject differentiation is accomplished. The lumbar spine is clearly visible because it is relatively opaque in relation to the adjacent relatively radiolucent tissues.

When significant tissue differentiation does not exist, contrast must be augmented by means of radiopaque or radiolucent contrast agents. One has only to review a survey film of the abdomen to reacquaint oneself with the inherent technical limitations. Although a survey film of the abdomen will demonstrate the bones of the lower ribs, the spine, and the pelvis, only an outline of the liver and spleen can be seen. Both kidneys are represented as faint shadows. Gaseous pockets distributed throughout the abdomen can be imaginatively linked together to form and outline of the colon, or the stomach. The lateral margins of the psoas muscles are generally prominently seen. The stomach, large bowel, small bowel, biliary system, urinary system, arterial and venous systems, all require a contrast agent (radiopaque or radiolucent) to be visualized on a radiograph.

Kilovoltage and Contrast Agents

There are some general rules, as well as the usual exceptions, about the proper kilovoltage range to use with specific contrast materials. With the exception of studies requiring barium sulphate, almost all contrast

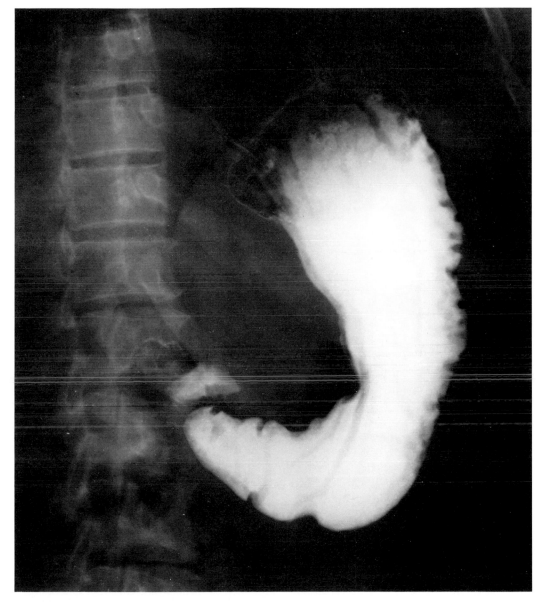

Fig. 6-5. Exposure at 150 kVp of a barium-filled stomach. Peristaltic motion is virtually eliminated with high kilovoltage technic, 125 to 150 kVp, because of the short exposure times required. Note the superb detail of the rugal folds of the stomach because of the penetrating effect of 150 kVp.

studies use low or moderate kilovoltage values. Adequate penetration of the barium-filled organs for a "see-through" effect is achieved with high kilovoltage technics. Shorter exposure times help to hold patient or organ motion blur to a minimum. Higher kilovoltage values can be utilized if there is a direct administration of the opaque into the area under study. For example, high kilovoltage produces excellent contrast studies of the stomach, small bowel, or the colon (Fig. 6-5). Studies of the upper GI tract require the drinking of barium, whereas the colon receives barium in retrograde fashion.

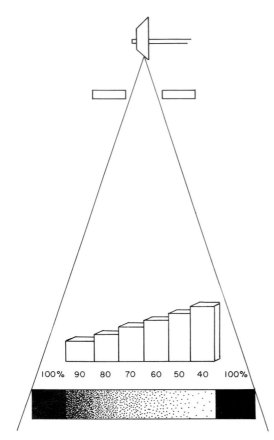

Fig. 6-6. Primary beam attenuation. A step wedge of aluminum is used to demonstrate the attenuating effect of different density levels. There is very little natural contrast differentiation within the body. Contrast agents must be used to augment radiographic contrast. Radiopaque agents attenuate the *primary* beam, resulting in good subject contrast.

A low to moderate kilovoltage value should be used when there is an indirect routing of the opaque prior to reaching the organ under study. If the urinary system or biliary tract is being examined by an intravenous technic, there is a decrease in the density of the opaque because of the indirect routing of the contrast agent, with subsequent dilution of its density. Low kilovoltage values are needed to achieve a high contrast effect.

There is considerable difference in the appearance of a common duct as visualized by intravenous cholangiography and as seen by direct-injection postoperative or T-tube cholangiography. The intravenous study depends on the functioning of the liver for deliverance of the contrast agent to the biliary system. The opaque shadow is generally weak in nature, and every technical effort must be utilized to effect maximum contrast enhancement of the image. The postoperative direct-injection cholangiogram exhibits extremely high contrast as a result of the direct injection of undiluted contrast material into the biliary system. The direct-injection technic is more satisfactory from a photographic standpoint, but can be totally unsatisfactory from a diagnostic standpoint. A nonopaque biliary calculi can be "lost" in an extremely dense contrast agent. Moderate or high kilovoltage values are therefore required with the direct-injection technic to penetrate the dense opacified biliary ductal system for residual radiolucent biliary calculi. Moderately high or high kilovoltage levels almost completely destroy the contrast level of the typical intravenous cholangiogram.

The size or caliber of a biliary duct can also have a definite influence on kilovoltage level. In performing operative cholangiography, it is not uncommon to see large dilated biliary ducts that require large amounts of opaque. *A dilated biliary ductal system must be adequately penetrated by kilovoltage when residual nonopaque biliary calculi are suspected* (Fig. 6-7).

The major exception to the direct-injection, high kilovoltage rule is the angiogram, often made after a direct injection of contrast into a vessel or artery. Since the opaque is rapidly diluted by blood, small vessels exhibit little or not subject contrast. Moderate kilovoltage, high milliamperes, and short exposure times are required.

Studies utilizing air or gas as a contrast agent require relatively low to moderate kilovoltage values. These studies include cerebral ventriculography, contrast introduction via a trephine opening in the skull, pneumoencephalography, introduction of air via the lumbar subarachnoid space, air contrast studies of the colon, and presacral air studies.

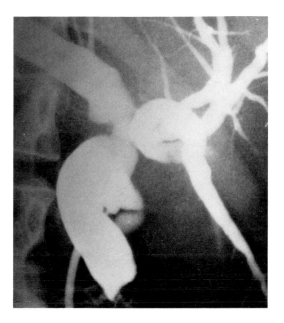 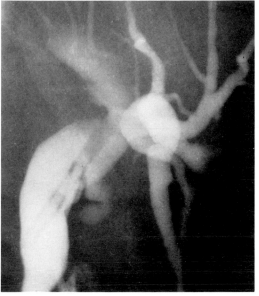

Fig. 6-7. Adequate penetration of an opaque-filled biliary ductal system. Moderately high kilovoltage values are needed to penetrate a dilated opaque-filled biliary ductal system. The radiograph on the left was exposed at 70 kVp and is underpenetrated. Residual radiolucent biliary calculi might not be demonstrated with this technic. The radiograph on the right was made with 90 kVp. Adequate penetration assures the radiologist of the absence of radiolucent filling defects.

THE USE OF HIGH KILOVOLTAGE WITH CONVENTIONAL TECHNICS

High kilovoltage values have been fairly well accepted for radiography of the chest and barium-filled organs. Some unusual uses for moderately high or high penetration technics are suggested in the following.

Chest Radiography

Some departments utilize 65 to 70 kVp non-grid chest technic for chest radiography, whereas others insist on high kilovoltage (125 to 150 kVp) grid or air-gap technics. This high kilovoltage range adequately penetrates ("sees behind") the heart without undue prolongation of the exposure time. High kilovoltage can be extremely helpful if one is interested in arresting the motion of the heart valves when a cardiac chamber analysis is being performed.[4]

Overpenetrated Bucky Films of the Chest. Years ago when all chest radiographs were made with the use of non-Bucky on nongrid technics, a request for a Bucky film of the chest was rarely misinterpreted. An overpenetrated film of the contents of the thoracic cage was taken. The patient was placed in the AP supine projection, and a dark or overpenetrated film was made with a dorsal spine technic. Since dorsal spine technics are based on the visualization of bone detail, 1 second or longer exposures are used to blur the vessels of the lung, and two technical errors are accordingly made: (1) inadequate penetration, and (2) lengthy exposures, or at least exposures longer than those acceptable for chest radiographs.

If a Bucky-type radiograph is desired, high kilovoltage ranges should be utilized in a supine or erect position.

Lateral Shoulder and Humerus Radiography

Dislocation or fracture of the humerus is an acute orthopedic emergency, and no one is more aware of this than the injured patient.

The pain with this type of injury can be excruciating, and many patients faint, or at least feel faint, during the radiographic procedure. Great emphasis has been placed on the use of patient breathing during the lateral projection of the humerus. An exposure is made with a low milliampere value, a relatively long time, and a moderate kilovoltage. The patient is asked to breathe slowly and rhythmically to blur the markings of the lungs and the ribs. Often a patient is in no condition to cooperate, and the result is a blurring of the humerus as well as the contents of the thoracic cage. An alternate technic (40 inch focus-film distance and maximum inspiration) requires the use of approximately 100 kVp. The x-ray beam is collimated to film size, with starting factors of 200 mA at $\frac{1}{10}$ of a second. Small patients require 90 kVp, medium patients 100 kVp, and large patients 110 kVp.

The patient is positioned with the injured arm at the side and the opposite arm elevated to avoid superimposition, he is instructed to take in a deep breath, as with a lateral chest radiograph, and an exposure is made. Several advantages result: (1) the study can be completed quickly, (2) rarely does a radiograph exhibit motion of a part or the patient, (3) with the use of the high kilovoltage, low milliampere technic, the lung fields do not darken or blacken, and (4) a higher kilovoltage value provides a greater margin for error.

The technic is basically a lateral chest projection at a 40 inch FFD with a high kilovoltage technic.

Nasal Radiography, Superior-Inferior Projection

Many views are taken of the nasal bones, including the Waters' or stereo-Waters' projection, lateral soft-tissue films, and the often neglected occlusal film in the superior-inferior projection (Fig. 6-8, *left*). With the advent of automatic processing, multiple radiographs can be made of the nose and viewed in minutes. Occlusal film must still be hand-processed or taped to a lead film and pulled through the automatic processor. There is, therefore, a perfect excuse for ignoring the making of the superior-inferior

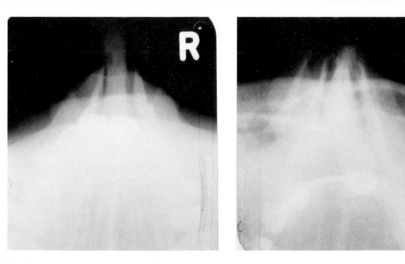

Fig. 6-8. Superior-inferior projection of the nasal bones. When a conventional superior-inferior projection of the nasal bones (*left*) cannot be obtained because of the small size of a nose or prominent central incisors, a film can be made with the chin intentionally depressed to superimpose the nasal bones on the frontal segment of the skull. A high kilovoltage technic (100 kVp or more) is used. The central ray enters the skull slightly behind the frontal sinus area, and the nasal bones are demonstrated through the skull itself (*right*).

occlusal film. Since some nonscreen film can be processed in an automatic processor, a corner of a 5″ × 7″ prepacked folder can be positioned in the mouth for an occlusal-type radiograph of the nasal bones. A lead mask covers one half of the film. The other half of the film can be used for a second exposure of the nasal bones. This may seem like a waste of film, but a 5″ × 7″ nonscreen ready-pack film is relatively inexpensive as opposed to the cost of the time required to hand-process an occlusal film adequately. *Be certain that the nonscreen product used for this technic can be processed in your particular automatic processing system.*

Despite the simplicity of the superior-inferior view, technical difficulties are encountered. For example, when the film is properly placed in the mouth with the patient seated erect, the film should be parallel to the floor. A perpendicular beam technic is used, with the central ray entering at the top of the nose anterior to the frontal sinuses (Fig. 6-9A). Prominent central incisors will invariably be superimposed upon the nose, preventing the visualization of the nasal bones. The experienced technologist soon learns to evaluate a patient to determine whether it will be possible to obtain an adequate superior-inferior radiograph. A special positioning modification must be made and high kilovoltage used with many patients. The patient is seated erect and is made to lean forward. This seating arrangement, with the moving forward of the head, avoids direct exposure to the gonadal area. The nonscreen or occlusal film is placed in the mouth and a positioning change is made. The patient's forehead is moved forward, and the chin is depressed. The frontal segment of the cranium is deliberately superimposed on the nose (Fig. 6-9B and C). The use of 100 kVp or better permits penetration of the forehead with superimposition of the nasal bones on the relatively homogeneous bones of the forehead. The central incisors are shifted posterior, due to the anterior shifting of the forehead, and do not superimpose on nasal bones (Fig. 6-8, *right*). Even if a patient

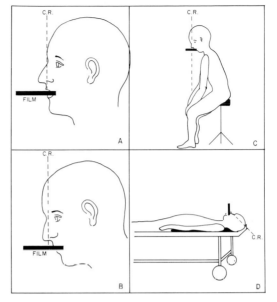

Fig. 6-9. Nasal radiography, positioning technic. A conventional position (A) is compared with a unique position (B) for radiography of the nasal bones in the superior-inferior projection. A high kilovoltage technic (100 kVp or better) is suggested for patients with prominent central incisors or an unusually small nose. The chin is depressed, and the central ray (C.R.) enters the forehead, superimposing the nasal bones on the anterior skull (B). The patient can be seated erect (C) and tilted forward to avoid direct gonadal dosage; a lead apron is a *must*. Or he can be examined in the supine position (D).

has a relatively small nose with prominent protruding central incisors, the nasal bones can be demonstrated. Great care should be taken with positioning, for it is difficult to justify repeated exposures to radiation in the vicinity of the eyes. If it is impossible to place the patient in an erect position, the recumbent position can be utilized, with the tube-angle technic avoiding gonadal dosage (Fig. 6-9D). *Regardless of technic, processor, or patient difficulties, this view should be attempted because of its diagnostic importance.*

Localization of Opaque Foreign Bodies

High kilovoltage values for small objects, such as straight pins, nails, etc., can be tolerated. It is advisable to use high milliamperes, high kilovoltage, and a rapid exposure time to "capture" linear metallic objects that may be in motion. Straight pins can produce a blurred, streak-like artifact with a relatively long ($\frac{1}{4}$ to $\frac{1}{2}$ second) exposure time. The use of a moderate kilovoltage, non-Bucky technic for evaluation of an opaque foreign body in an infant or baby is important. Films are frequently taken in the wrong sequence, and an opaque foreign body is not seen.

It has been the experience of this author on several occasions to examine the abdomen of an infant for a foreign body and to find no evidence of a foreign body on the finished film. Minutes later, radiographs of the chest were exposed, to be certain that the foreign body was not lodged in the trachea or the esophagus. Again, there was no evidence of a foreign body. A third set of films was taken of the abdomen, which this time demonstrated a foreign body, usually a coin. Such a sequence means that the coin was lodged somewhere in the neck or the esophagus when the initial study of the abdomen was made, and then in the excitement that ensued while the first films were being processed, the coin passed into the abdomen.

A small child should be examined from the nose to the anus, and in that order, in case a foreign body is lodged in the nose, in the throat, in the chest, in the trachea or esophagus, in the abdomen, the rectum or vagina. A simple technic (1 exposure) "sees through" both the thick and thin areas of the body with a moderate kVp (85-90) technic. The child should be completely undressed to make certain that the coin or pin is not within or on its clothing, and positioned diagonally on a 14" × 17" cassette. The head is turned to one side to demonstrate the nose, face, and air structures of the neck. A film is exposed of the entire body (400 mA, $\frac{1}{120}$ of a second, approximately 85 kVp). Linear metallic foreign bodies are "captured" if in motion, and all natural body cavities are examined for the presence or absence of an opaque foreign body.

Foreign Body Localization in the Eye

One has only to attempt to visualize an opaque foreign body in the eye to realize how poor most screen cleaning habits are. Technologists never seem to notice the dust that produces metallic-like artifacts on a finished radiograph until an attempt is made to radiograph an eye for a foreign body.

A simple method using intensifying screens is to take 2 radiographs in a shallow Water's position, one with the eyes looking to the left, and a second with the eyes looking to the right. If a foreign body in the right eye were to be anterior to the orbit, it moves laterally when the eyes move to the right, and it moves medially if posterior. The same effect would be noted in the left eye when it is looking to the left. A lateral radiograph of the eye could be taken with the patient first looking up at the ceiling and then looking down at the floor. When the eye is turned upward, the foreign body will move up if anterior, and down if posterior; and when the eye is moved downward, the foreign body will move down if anterior and up if posterior. If it is determined that a foreign body is present in the eye, and a method for accurate localization is required, a grid or Bucky nonscreen film in a direct-exposure holder with 100 kVp (200 mA at $\frac{1}{2}$ second) in the lateral projection can be used. There is no blackening of the air-filled cavities of the sinuses, and the foreign body is easily seen on the finished radiograph. The relationship of the foreign body to a localization device is obvious, since the film is of a lighter density. Measurement technics are quite easily accomplished.

SOFT-TISSUE TECHNIC, INCLUDING MAMMOGRAPHY

The use of soft-tissue technic radiography seems to be confined at the present time to nonscreen radiographs of the extremities or radiography of the breast.[17] Several technics can be used to evaluate the breast adequately,[19] and there are definite differences of opinion about the types of film, focus-film distances, focal spot size, etc., to be used. Yet some general rules are shared even by those advocating diametrically opposed technics. These include: (1) conventional screen films in nonscreen holders are not adequate, (2) extremely low kilovoltage values are needed for contrast enhancement of soft-tissue structures or minute calcific densities within the breast, (3) relatively high mAs technics are required to produce adequate film blackening, and (4) a nonscreen medical film or a nonscreen industrial film must be used.

The type of film required seems to be the major subject of disagreement in technic. Dr. J. Gershon-Cohen has for years advocated the use of a nonscreen medical film.[9] Dr. Robert Egan has insisted on the use of a fine-grain industrial-type direct-exposure film,[5] whereas Stanton and his colleagues used a slightly faster industrial film.[18,19] "Sandwich" packs of industrial films—2 films, one of faster speed than the other— are commercially available and simultaneously yield 2 radiographs of each projection of the breast. One is an extremely dense film requiring a hot or bright light for interpretation; the second film is considerably lighter in density, demonstrating thinner areas of the breast. Several classic references with technical emphasis are listed.[3,5,9,18] It is the opinion of this author that almost any technic is adequate if strict adherence to the rules of that technic are observed.

New units are available having molybdenum target radiographic tubes and molybdenum filters to produce a monochromatic beam for contrast enhancement of the soft tissues of the breast. Although the molybdenum target and molybdenum filter have been quite effective in improving mammographic technic, the use of a compression device in conjunction with this system (a device that clamps the breast gently but firmly, reducing the wedge shape of the breast to a more even overall density) has effected a major improvement in these studies.

REFERENCES

1. Cahoon, J. B., Jr.: Radiographic technique: its origin, concept, practical application and evaluation of radiation dosage. X-ray Techn., 32(4):354–364, January 1961.
2. Cullinan, J. E., Jr.: Choleroentgenography— history and development of current biliary opacification techniques. Radiol. Techn., 40(4):226–235, 1969.
3. Curcio, B.M.: Mammography and the radiologic technologist. Radiol. Techn., 38:143–151, November 1966.
4. Eastman, T. R.: Chest technique through body habitus. Radiol. Techn., 41(2):80–84, 1969.
5. Egan, R. L.: Experience with mammography in a tumor institution: Evaluation of 1,000 studies. Radiology, 75: 894–900, December 1960.
6. Fuchs, A. W.: The optimum kilovoltage technique in military roentgenography. Amer. J. Roentgen., 1(3): 358–365, September 1943.
7. ———: Principles of radiographic exposure and processing. Springfield, Ill., Charles C Thomas, 1958.
8. Funke, T.: Pegged kilovolt technique. Radiol. Techn. 37(4):202–213, 1966.
9. Gershon-Cohen, J., and Ingleby, H.: Carcinoma of the breast: Roentgenographic technic and diagnostic criteria. Radiology, 60:68–76, January 1953.
10. Koenig, G. F.: Potential. Radiol. Techn., 37:184–198, January 1966.
11. Lyons, N. J.: Optimum kilovoltage techniques. X-ray Techn., 33:398–401, May 1962.
12. McDaniel, C. T.: Tissue absorption theory applied to formulation of radiographic technique resulting in time as the variable factor. Radiol. Techn., 37(2):51–55, 1965.
13. O'Donnell, E. B.: Concerning high kilovoltage technique and accessories. Radiol. Techn., 36(2):71–75, 1964.
14. Ott, T. T.: Extending the range of medical radiography to 150 kilovolts. X-ray Techn., 34:335–343, May 1963.
15. Report of the Medical X-Ray Advisory Committee on Public Health Considerations in Medical Diagnostic Radiology (X-Rays) U.S. Department of Health, Education, and Welfare; Washington, D.C., U.S. Government Printing Office, October 1967.
16. Ross, J. A., and Galloway, R. W.: A Handbook of Radiography. ed. 3. Philadelphia, J. B. Lippincott, 1963.
17. Seemann, H. E., and Lubberts, G.: Films for soft tissue radiography. Med. Radiogr. Photog., 41(1):18–20, 1965.

18. Stanton, L., and Lightfoot, D. A.: The selection of optimum mammography technic. Radiology, *83*(3): 442–454, September 1964.
19. Stanton, L., Lightfoot, D. A., Boyle, J. J., Jr., and Cullinan, J. E.: Physical aspects of breast radiography. Radiology, *81*:1–16, July 1963.
20. Variability in the Automatic Processing of Medical X-Ray Film. U.S. Department of Health, Education, and Welfare, No. BRH/DEP 70–13. Washington, D.C., U.S. Government Printing Office, June 1970.

7. Mobile or Bedside Radiography

With the advent of modern high-output radiographic units, bedside radiography has finally come into its own. Battery-operated or condenser-discharge mobile units using self-contained or conventional power sources are available, as well as mobile units which use 220 volt power lines duplicating the output of conventional radiographic equipment.

The use of lightweight mobile units will be stressed in this chapter. Most smaller bedside units have a limited output (15 to 30 mA, 85 to 90 kVp). Technical problems pertaining to all mobile units but specific to low mA, moderate kVp units will be discussed.

TYPES OF MOBILE RADIOGRAPHIC UNITS

A listing of available portable, mobile, or bedside equipment would include the following:

1. **Fifteen to 30 mA units with an 85 to 90 kVp limitation.** These units are used with 110 volt circuits for every type of mobile radiographic examination, anything from radiography of the chest of a newborn infant to radiography of the hip in the lateral projection. Low-output equipment is inadequate for many types of radiographic examinations. For example, many of these units have severe timer limitations, $\frac{1}{10}$ of a second being the shortest exposure time available to the technologist. Since the heart of an infant can beat as rapidly as 140 times per second, it is difficult to arrest motion, even with a high-speed impulse timer ($\frac{1}{120}$ of a second). If the 140 pulsations of the heart per second are divided by $\frac{1}{10}$ of a second (14 pulsations per $\frac{1}{10}$ second), it is easy to see how inadequate a low-output mobile unit can be for radiography of the infant chest. The use of this unit in the 85 kVp range for a lateral hip projection is equally disturbing, for patient thickness through this area can exceed 25 cm, necessitating excessively long exposure times. When a long exposure time is combined with a low ratio grid and a poor collimating device, it is remarkable that satisfactory radiographs can be obtained.

A definite advantage of a self-rectified unit is that it is relatively easy to move from area to area, since its weight is less than half that of larger units. They are generally quite reasonable in price and use a standard power supply.

A major disadvantage is the relatively long exposure times required for grid-type radiographs. Motion is a chronic problem even in attempting radiographs of the supine chest in the range of $\frac{1}{10}$ to $\frac{3}{10}$ of a second.

2. **High-output mobile units requiring a 220 volt power supply.** These units are not in common use, for the cost of rewiring an entire hospital with 220 volt lines of 50 amp or greater capacity is prohibitive.[4] Hundreds of thousands of dollars would be needed to wire properly each patient room, a sum which scarcely seems practical in view of the fact that many 220 volt mobile units can be purchased for under $10,000.

If one is fortunate enough to have a high-output unit and adequate wiring, it can be used in many ways, such as:

1. A mobile unit[4] can be wired to an existing cassette or serial film changer for vascular

127

emergencies (cerebral, thoracic, or abdominal angiography). A second cassette changer can be used with a 200 mA mobile unit for lateral cerebral angiography. The mobile unit can be wired synchronously with the angiographic unit for simultaneous biplane cerebral angiography. A mobile unit can be used effectively with a cassette changer in the accident ward or the operating room.

2. A heavy-duty mobile x-ray unit can be substituted for a conventional unit. During peak work periods a mobile unit can be used for routine diagnostic procedures in a superficial or deep therapy room, if the therapy work schedule permits.

3. Mobile units can be adapted for breast radiography. If a limited number of mammographic examinations are performed in a department, and it is difficult to justify the purchasing of a mammographic unit, a special cone or collimator can be easily adapted to a mobile unit for mammographic technics.

4. A mobile unit can act as a supplemental tube for cross-table projections if the tube on the conventional unit is mounted in a fixed position.

The major advantage of the full-wave rectified mobile is the increased radiation output, making it possible to perform examinations of departmental quality at the bedside. A distinct disadvantage is the weight of the unit as compared with that of the lower-output equipment. These units are difficult to maneuver, particularly in large hospitals with inclined ramps and lengthy tunnels. It can be even more difficult to stop a unit of this type on an inclined plane. The installation of carpeting in many hospitals has made it harder for the average female technologist to move a unit of this weight. Some of the newer, heavy-duty mobile units come equipped with an automatic drive or power assist. When such a unit has a motor which operates on a rechargeable battery, moving or guiding the unit in a small, crowded patient room or on an extended run through a large hospital becomes a simple matter.

On occasion it is desirable to use a heavy-duty mobile unit in a special area such as the accident ward, intensive care unit, operating room, recovery ward, or pediatric nursery. For an employee chest survey, heavy-duty lines can be installed even in an area remote from the main department. The use of a high-output mobile unit is restricted only by the lack of ingenuity of the operator.

The use of this unit at a remote location requires special wiring. Many hospital sections have individual air conditioners which use separate 220 volt sources. The plug and receptacle of the air conditioning unit can be adapted to match the plug of the mobile radiographic unit. When the mobile unit is used at this location, the temporary disconnection of the air conditioner plug gives access to wiring that provides the mobile unit with an independent 220 volt power supply.

When high-power outlets are installed at random locations throughout the hospital, long extension cords are required. Because of power loss in the cable, the output decreases as the length of the extension cord increases. The handling and storage of a long cable can be a nuisance. The longer the cable, the more dirt it collects; in being handled, soil is transferred from the cable to uniforms, stockings, shoes, etc.

3. Capacitor-Discharge Unit. A distinct advantage of the capacitor or condenser-discharge unit is that it requires no special wiring since it uses a standard 110 volt source to build up a charge of electrical current. After an adequate charge period, an exposure can be made with a significant increase in radiation output, as compared with that of the self-rectified units. A capacitor-discharge mobile is generally less expensive than a heavy-duty (220 volt) mobile and approaches a self-rectified unit in size and weight.

When the capacitor-discharge unit is used with a standard FFD, it is sometimes impossible, due to the limited capacity of some of these units, to radiograph the heavier segments of the body. When more film blackening is needed, it is recommended that the FFD be shortened; that is, the tube should be placed closer to the part even though many of these units have a relatively large focal spot. Grid focus can be a problem with a short focus-film distance. If the maximum output of the unit is reached with a single exposure and film density will not be satisfactory, it is necessary with some of the limited-capacity units to make a double exposure, using the same cassette. Unfortunately, the unit must be recharged between the 1st and 2nd exposures.[5]

4. **Battery-powered Mobile Unit.** Mobile radiographic units using nickel cadmium cell groups can be recharged from a standard 115 volt, 5 amp outlet.[5] The battery package has a capacity of 10,000 mAs per charge (100 mA at 110 kVp, using a rotating anode tube of 3,000 rpm). These units have significantly more output than a capacitor-discharge unit, and are priced in the range of the full-wave rectified mobile. Since they carry their own source of power (nickel-cadmium cell groups), they weigh more than the capacitor-discharge units. The manufacturer of this unit claims that radiation output is similar to that of a 3-phase generator with a 12-pulse rectification system (100 mA output is said to be equivalent to 150 mA from a 2-pulse generating system). Since there is no surge or falling off of output, the tube does not have to be protected with a large focal spot as does the capacitor-discharge tube. Using a 1 mm focal spot with a standard FFD will help the technologist to avoid image unsharpness and grid focus difficulties.

When fast screens and fast film are used with this mobile, it is capable of almost any technical exposure required for bedside radiography.

Future Use of Cordless Equipment (Nickel-Cadmium Cells). Nickel-cadmium cells, although relatively new to the x-ray field, are more durable and lighter in weight than lead-acid batteries, and do not deteriorate as rapidly with age. Such cordless equipment may become the pattern for permanent installations in the future since the battery package will not have to be wheeled around the hospital. Increasing the size of the energy storage unit could lead to higher

ratings. The present cordless unit has a rating equal to that of many medium-power permanent installations.[5]

COMPARISON OF MOBILE EQUIPMENT

Some important features to be considered in contemplating the purchase of a mobile unit are:

1. "Dead man" type of brake with positive locking action. When pressure is removed from the bar that is used to push the mobile, a "dead man" brake takes hold.[4]

2. Actual physical size. The vertical dimension of the unit must be such that it can pass through the opening of the lowest door in the building where the unit is to be used. The maximum width is determined by the width of the narrowest door. A small unit is invaluable in a room with closely spaced beds.

3. Tube arm height and mobility. It must be possible to achieve a 40 inch FFD. The transverse tube arm should extend sufficiently so that it can be used on the opposite side of the operating field for such procedures as cholangiography. The unit should be capable of being parked on the left side of the patient, and the tube arm extended to the right upper quadrant for adequate centering, permitting the tube angulation technic shown in Chapter 2. The x-ray tube should be capable of being locked in a park position so that the tube is not damaged in transit. All locking mechanisms, as well as counter balance adjustments, must be carefully maintained. The tube, collimator, and tube arm must be *failsafe*; that is, it should not be possible for them suddenly to plummet to the floor, risking injury to the patient.

4. Proper grounding of the equipment. This is mandatory. The purchase of a full-wave rectified mobile radiographic unit is restricted to a hospital or clinic where adequate wiring already exists. If the wiring of a facility (220 volts) is out of the question, then a self-rectified or capacitor-discharge unit utilizing a conventional power source

(110 volts) should be considered. A self-rectified unit with low mA and moderate kVp output is quite adequate for radiography of the chest or most routine orthopedic studies, although adequate penetration of the hip in the lateral projection can be difficult. Films of the pelvis and abdomen require an exposure of 1 to 3 seconds with fast-speed intensifying screens and a low ratio grid.

With a capacitor-discharge unit using a 110 volt line, good-quality chest and extremity radiographs are possible, but some capacitor-discharge units are somewhat limited for thicker body parts. The battery-powered unit seems to meet almost all criteria for a high-output mobile radiographic unit. It weighs less than many full-wave rectified units, and delivers a high radiation output without the power loss resulting from the use of long extension cords.

LOGISTIC PROBLEMS WITH MOBILE UNITS

Before a high-output mobile unit is put into service, definite criteria for its usage must be established. When the medical and surgical staff of a hospital discover that radiographs of high quality can be made at the bedside, there is an alarming increase in the number of requests for this service. The use of a mobile unit to supplement an inadequate escort service can never be condoned, although critically ill patients who cannot tolerate a lengthy trip to the radiology department can often be better served by a bedside examination.

Physicians and nursing personnel should be made aware of the limitations of even the finest bedside study. A special effort must be made to acquaint the orthopedic surgeon with his responsibility to the radiology service. Invariably, a patient in a complex traction device is situated in a room in a manner that prevents proper radiographic positioning. For example, a patient with a fracture of the right femur is placed with his left side against a wall. A true lateral projection of the entire femur cannot be secured. Some traction set-ups (with the leg elevated) virtually eliminate the use of a focused grid because of the shortened FFD. Often an AP projection of the femur will be overexposed (due to shortened FFD), and the lateral projection will be underexposed (due to an increased FFD) because of a positioning difficulty.

ASPECTS OF BEDSIDE RADIOGRAPHY TECHNIC

Variation in Focal Film Distance

Since an injured extremity can be placed in any number of positions in a traction device, technologists encounter both shortened and/or increased focal object distances. A 30 inch FFD could conceivably be reduced to a 15 inch FFD for an AP projection. A 60 inch FFD could be required for a lateral projection to overcome a positioning difficulty. A 1 second exposure at a 30 inch FFD would have to be lowered to $\frac{1}{4}$ second (15 inch FFD) for the AP film, and increased to 4 seconds (60 inch FFD) for the lateral film. These adjustments are determined by the *inverse square law:* the intensity of the x-ray beam varies inversely with the square of the focal film distance. Seldom are distance changes of this magnitude encountered; therefore there is a tendency to ignore minor variations in distance. Minor variations can have a major effect on film density. A yardstick or tape measure for measuring distance accurately is a prerequisite for bedside radiography.

In the following, some variations in distance and the effect on exposure length are compared with a 1 second exposure value at a 30 inch FFD.

New Distance	New Exposure Value
25″ FFD (−5 inches)	$\frac{7}{10}$ second (−$\frac{3}{10}$ second)
36″ FFD (+6 inches)	$1\frac{2}{5}$ second (+$\frac{2}{5}$ second)
40″ FFD (+10 inches)	$1\frac{3}{4}$ second (+$\frac{3}{4}$ second)
48″ FFD (+18 inches)	$2\frac{1}{2}$ second (+$1\frac{1}{2}$ second)

A $33\frac{1}{3}$ per cent increase in distance (from 30 inches to 40 inches FFD) necessitates a 75 per cent increase in exposure time (from 1 second to $1\frac{3}{4}$ seconds).

Line Voltage Compensation

The line voltage compensator should be adjusted prior to every radiographic exposure. Many hospitals are gigantic complexes composed of new and aging structures, and although electrical power may be adequate in one area, it does not follow that an ample supply of power is available to the entire facility. As more complex electronic devices become more commonplace, greater demands will be placed on existing power sources. *The main voltage correction must be accomplished before each exposure at the bedside.* Vennes and Watson recommend the lowering of the mA value with an increase in exposure length to overcome inadequate circuitry.[8]

Some General Considerations Concerning Bedside Radiography

Always report to the nurse in charge prior to initiating a bedside study. A technologist can easily be deceived about the condition of a patient. The patient may appear to be resting comfortably and show little evidence of the seriousness of his condition,[8] which could have worsened in the time interval between the ordering of the radiograph and the arrival of the bedside equipment.

Every effort should be made to understand the condition of the patient with an orthopedic problem. Specific information must be obtained before moving or lifting the patient in traction for cassette placement.

Why are frequent re-examinations ordered for the same patient? The condition of a patient can change abruptly, as well as drastically, in a short period without an obvious change in appearance. Patients with chest difficulties can be studied repeatedly to learn whether a collapsed lung is reexpanding, or whether thoracentesis (the removal of fluid from the chest) has been effective, or whether there is residual fluid. Fluid can reaccumulate; therefore the sequence in which the radiographs were taken must be accurately documented.

Films taken during a procedure such as reduction of a fracture of a wrist or a hip pinning must be marked with a time sequence, such as Set No. 1, Set No. 2, etc. The technologist does not have the right to destroy seemingly useless radiographs taken during a prolonged procedure. *Proper marking sequence must be used with multiple studies of the same patient.* Each radiograph must be marked with the proper date and the sequence of exposures recorded directly on the film. The actual time at which the examination was made should always be noted on the request.

Patients with bone injuries are examined one or more times daily as traction manipulations are made. AP and lateral projections are mandatory (Fig. 7-1). *Accurate tube-film alignment is crucial to patient care* (Fig. 7-2).

The use of proper lead markers to designate the right or left side cannot be overstressed. Supine and erect markers should be used to help the radiologist reach a diagnosis. If the decubitus position is attempted, proper markings should also be used with the film. A change in procedure or technic should be recorded on the x-ray request. Short simple sentences should be used, for x-ray requests and films are legal documents. Current ver-

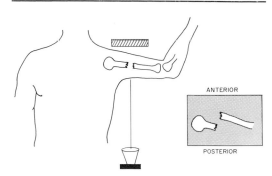

Fig. 7-1. The value of multiple projections, one at a right angle to the other. Conventional radiographic projections are often ignored during a bedside assignment. The value of a 2nd projection taken at a right angle to the initial film cannot be overstressed. A film of the humerus in the AP projection can demonstrate perfect alignment of the fragments of the humerus, but a lateral or axial projection shows that there is marked anterior displacement of the distal fragment.

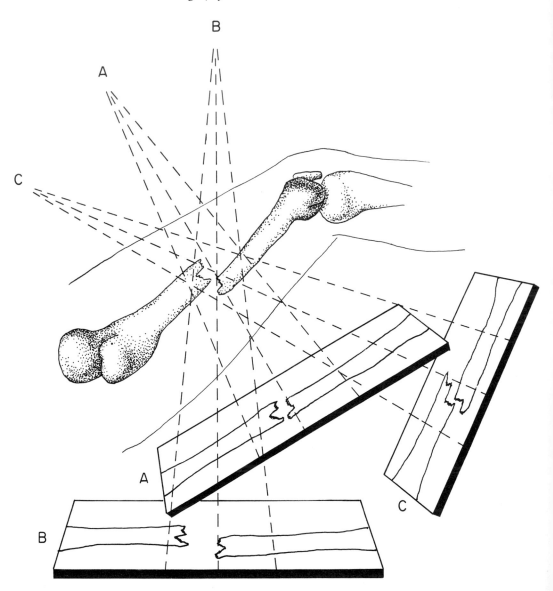

Fig. 7-2. Tube, film, and object alignment. Careful attention must be given to tube-film-object alignment in using a bedside unit. Tube and film A are in proper alignment, and a slight separation of the fragments of the femur is demonstrated. Tube B is so placed that the resulting radiograph (B) gives the illusion of fragment separation. The misalignment of tube C gives the illusion of fragment overlapping.

The technologist should study any previous films or technic records that are available prior to a bedside examination of a fractured extremity. At least 1 joint must be included on each projection.

nacular should be avoided, for the technologist may be called upon to explain his notations in a court of law.

Repeat Bedside Examinations

There is a tendency to accept bedside films of poor quality if for no other reason than that they are portable radiographs.

The major technical complaint concerning bedside studies is that films are frequently overexposed. Careful attention must be paid to the differences between timer increments. In the lower range ($\frac{1}{10}$ to $\frac{3}{10}$ second) a 1-step variation will produce a 50 to 100 per cent change in film density. A 300 per cent increase in film blackening is produced by changing from $\frac{1}{10}$ to $\frac{3}{10}$ of a second (Fig. 7-3).

The technologist must always be aware of and concerned with the rating limitations of his bedside units. Most low-output machines use a stationary anode tube with severe rating restrictions. The technologist must be knowledgeable enough to be able to refuse to make an exposure that would endanger his equipment.

The use of any type of positioning aid (foam rubber, restraining bands, etc.) should be encouraged.

On occasion, when an unsatisfactory radiograph has been made as a result of overexposure or underexposure, breathing, etc., technologists are asked to repeat the study. A second attempt is made, but an equally dismal film emerges, perhaps a third attempt and a fourth attempt follow, depending upon the need for the study and the patient's condition. Finally, a suitable radiograph is obtained for interpretation, and the initial films are discarded. *Throwing away such films is extremely bad practice. No radiographic film should be thrown away without an evaluation by a quality-control technologist or a radiologist.* The poor-quality films establish the fact that an obvious effort was made to achieve a suitable radiograph. The physician can mention in his report that multiple attempts were made to examine this patient to no avail, and, last but most important, a subtle or questionable change in the final radiograph could possibly be substantiated or ignored when compared with one of the poorer-quality films.

Bedside Radiographic Factors Versus Conventional Factors

It is almost impossible to measure standard radiographic technics against mobile technics. See Table 7-1.

A specific technic chart must be fabricated for the mobile unit. It should contain **several basic considerations:**

1. A low-ratio grid should be considered because of grid focusing and grid cut-off difficulties. Low-ratio grids are compatible with the low output of the x-ray unit.

2. Fast-speed intensifying screens should be combined with faster x-ray films for shorter exposure times.

3. Radiographic density is increased as a result of the larger-size cones in use; since more patient area is exposed, more scattered radiation is generated.

4. Radiation output is considerably increased as the focus-film distance is lowered from the more conventional 40 inches to 30 inches.

5. An optimum kilovoltage technic is desirable, with emphasis on a moderately high kilovoltage value (85 kVp), whenever possible ensuring adequate penetration of the part under study, increased latitude, and the shortest possible exposure time.

Grid versus Non-grid Technic

Many areas of the body that require a grid or a Bucky for a conventional radiograph do not necessarily need a grid or Bucky for a bedside examination, since most bedside studies are made after an initial diagnosis has been determined. For example, a patient with a known fracture of the femur is placed in traction, and portable films are ordered to determine fracture alignment. A non-grid study of the femur, knee, or shoulder will result in an acceptable radiograph showing bone alignment. Grid radiographs require at least a tripling of the exposure factors as compared with radiographs made by non-grid technics.

LOW mA, MODERATE kVp OUTPUT, SELF-RECTIFIED UNIT

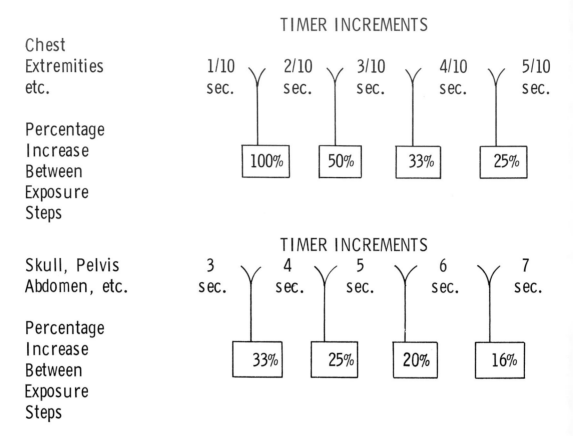

Fig. 7-3. Timer increments, low-output mobile equipment. A slight increase in the length of exposure during bedside chest or extremity radiography can be technically disastrous. If $\frac{1}{10}$ of a second is adequate for a particular projection and the timer is moved to the next increment ($\frac{2}{10}$ of a second), film blackening is increased by 100 per cent. The lower ranges are the most critical. A $\frac{1}{10}$ of a second increase in the $\frac{4}{10}$ of a second range adds only 25 per cent to radiographic density.

In the long range, one would have to increase an exposure value from 3 seconds to 6 seconds to achieve a 100 per cent increase in radiographic density.

Even if a radiograph can be taken with an adequate focus-film distance and an accurately focused grid, it is good technical practice to utilize non-grid exposures for fracture alignment. Some complicated fractures require continual adjustments to the traction and daily re-examination of the injured area. Every effort should be made to hold radiation dosage to a minimum. If a grid must be used, grid centering often can be more accurate with linear than with horizontal placement of a cassette (Fig. 7-4).

Supine versus Erect Radiography of the Chest

Many technologists examine the chest at bedside in the conventional PA erect position if the patient's condition permits. This means

Table 7-1. Typical Low-Output Self-Rectified Mobile Unit Compared With a Full-Wave Rectified Single-Phase Radiographic Unit

FACTOR COMPARED	SELF-RECTIFIED MOBILE UNIT	FULL-WAVE SINGLE-PHASE UNIT
Milliamperes	20 mA to 30 mA maximum	Up to 500 mA
Kilovoltage	85 kVp maximum	Up to 125 kVp
Timer (minimum length of exposure)	Hand timer ($\frac{1}{4}$ of a sec)	Impulse timer ($\frac{1}{120}$ of a sec)
Focus-film distance	30 inch FFD	40 inch FFD
Grid ratio	5 to 1	12 to 1 or greater
Primary ray collimation	Cone—generally one size	Light-beam collimator
Rectification system	Half-wave (every other impulse)	Full-wave rectification
Intensifying screens	Generally fast screens	Generally medium screens
Film	Regular or fast films	Generally regular film

that after the patient has been helped to sit up and place his legs over the side of the bed, he is required to hold a 14″ × 17″ cassette for the study. The erect PA film has a distinct diagnostic advantage over the AP erect or supine chest film, for it represents a radiograph that can be compared with conventional films.

Most patients are examined in the AP erect position. Since many hospital beds cannot be raised completely, a sort of AP apical lordotic position is obtained due to the angulation of the patient. This apical lordotic effect can be overcome by wedging a pillow in behind the cassette. The tube is first centered to the chest with the patient in a semi-erect position (Fig. 7-5A). The cassette is then used as a lever to pull the patient forward for the placement of the pillow, helping the patient to remain erect in the AP position (Fig. 7-5B). Technologists often place a mobile unit at the end of the bed, anywhere from 3 to 5 feet from the patient, without regard for focal object distance. As a radiographic tube is positioned farther from the film, a decrease in radiographic density occurs. Since most portable chest radiographs are taken at a 30 to 36 inch focus-film dis-

tance, the increasing of the focus-film distance to 5 feet (60 inches) requires an increase of 3 to 4 times in the mAs value.

Many radiographic departments examine all portable chests in the AP supine position and consider this position quite acceptable for several reasons: (1) The focus-film distance rarely changes if the x-ray tube is always raised to maximum height. An exception to this rule would be the elevation of a hospital bed on wooden blocks or the removal of the wheels from a bed for stabilization. (2) The patient is in a relatively comfortable position, and the film can be obtained without rotation of the patient. (3) The patient is not required to cooperate with the technologist by sitting erect or holding a cassette. Patient responsibility is limited to the holding of one's breath during the exposure.

Safety in Bedside Radiography

Many factors must be considered when safety during bedside radiography is discussed. Some of the more important considerations are:

1. The exposure control switch should be connected to a long cord (12 feet or greater)

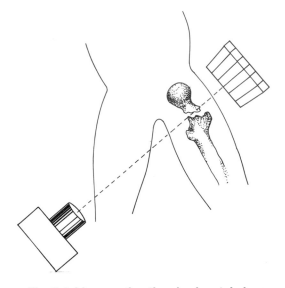

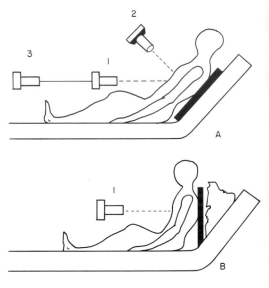

Fig. 7-4. Linear rather than horizontal placement of a grid. There can be a definite advantage, when radiographing the hip of extremely large or small patients, in placing the grid or grid cassette in a linear position rather than in a horizontal position.

The grid is positioned with the grid lines running perpendicular to the floor rather than horizontal to the floor. The x-ray tube can then be raised or lowered to the linear dimension of the grid to effect proper centering.

On an extremely thin patient the tube would be quite low, and on an extremely large patient the tube would be centered quite high in relationship to the cassette. This may not seem important, but if a cassette were to be used in the traditional horizontal position with a very thin patient, the central ray could be as much as 2 inches off the center of the grid. The reverse could happen with a rather large patient.

It is important to maintain right-angle intersection of the grid by the central ray and to see that the grid cassette runs parallel to the neck of the femur. The central ray must bisect the neck of the femur.

---

so that the technologist can avoid standing in the vicinity of the patient.[4]

2. The use of a lead apron by the technologist is an absolute *must* during bedside radiography. Often the lead apron is left in

Fig. 7-5. Focus-film distance errors. When an erect examination of the chest at the bedside is requested, several technical errors can be made. Assuming that the equipment is of relatively low output, a focus-film distance of 30 inches would be acceptable in Figure B_1.

When a patient is unable to sit completely erect, occasionally as in Figure A, the tube can be placed at a 30 inch focus-film distance (A_2) to assure proper film blackening. A common practice is to move the tube to the foot of the bed in the semierect or erect position, almost doubling the focus-film distance required for bedside radiography. (A_3) If the patient were to be left in the semierect position (A), an elongated, distorted, or apical lordotic type of radiograph would result with the tube left in the normal position (A_1). It is a simple matter to position the tube (A_1) with the cassette behind the patient and at the last instant, prior to the radiographic exposure, use the cassette as a lever to raise the patient into the upright position. A pillow must be wedged behind the cassette to assure the maintainence of this upright position (B).

With the exception of position A_2, any of the above tube positions would be accurate for the demonstration of an air-fluid level of the chest.

If the demonstration of an air-fluid level is requested, the central ray must run parallel to the floor, and every effort should be made to place the patient in a fully erect position at the appropriate focus-film distance.

Fig. 7-6. Panoramix beam. Note the umbrella-shaped beam which has a 270 degree angle of divergence. A cone-shaped target is mounted at the end of a 4½ inch extension of the x-ray tube. The electron stream strikes the apex of the target, which is made of copper with a tungsten apex. A focal spot size of 0.1 mm enables this tube to be used for air-gap technics during direct roentgen enlargement studies.

A beam limiting shield is shown in the lower part of the figure, which can be used to limit the scope of the x-ray beam.

Fig. 7-7. Panoramix Model MO-200. A mobile unit with a scissors extension arm for increased positioning flexibility. (Westinghouse Electric Corporation, X-Ray Division, Baltimore)

the x-ray department, for it is just too inconvenient to bring a heavy lead apron to a remote location.

3. A portable unit that is not considered *hazard-proof* should never be used in an area such as the operating room.[9] Whenever films are to be made in the operating room, the anesthesiologist should be consulted about the possible presence of explosive gasses.

4. A portable or bedside unit is rarely able to be used for fluoroscopic studies, for very few mobile units have adequate fluoroscopic collimation devices.

5. Additional radiographs should never be made at a remote location to avoid the possibility of a return trip for a repeat examination.

6. Gonadal shielding of the patient should be used whenever possible, especially when multiple follow-up studies are contemplated.

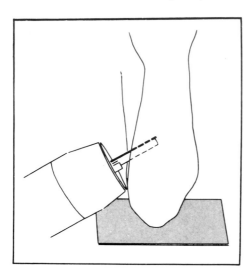

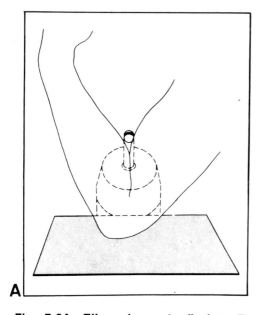

MOBILE PANORAMIX ROENTGENOGRAPHY

A unique radiographic tube originally designed for dental radiography can be used for direct roentgen enlargement technics (Figs. 7-6 and 7-7).[1,2,3,6,7] This unusual stationary anode tube has an effective focal spot of 0.1 mm, and can be positioned in any natural body cavity, producing an "inside-out" radiograph of unique value. The Panoramix (Westinghouse X-ray Corporation) has an almost body-section-like effect. It literally "lifts out" entire areas of the body, with accompanying image distortion and elongation. The unusual shape of the beam permits the use of distortion and elongation to free an area completely from superimposed details. (Figs. 7-8A and B, 7-9A and B, and 7-10A and B).

Although the maximum technical factors used with this unit (1 mA at 70 kVp for 1 second) seem rather low by conventional standards, caution must be used with this equipment because the x-ray source is often in direct contact with the patient. The users of this equipment are advised to read a report on the unit issued by the United States Department of Health, Education, and Welfare in March of 1970.[1]

Fig. 7-8A. Elbow in acute flexion. The Panoramix tube can be used to examine an elbow in acute flexion. A survey film can be made without extending the arm of the patient.

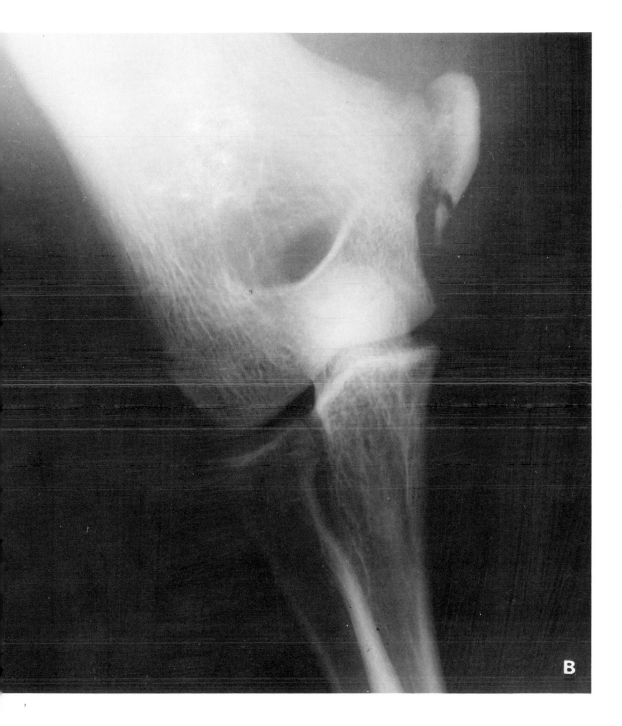

Fig. 7-8B. Enlargement study of the elbow in acute flexion. Several fractures are noted.

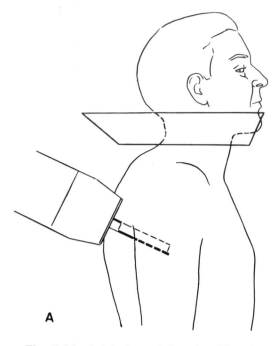

A

Fig. 7-9A. Axial view of the shoulder. The Panoramix tube can be used to obtain an inferior-superior view of the shoulder joint when abduction is impossible.

Fig. 7-9B. Enlargement of axial view of the shoulder joint with the arm strapped to the patient's chest.

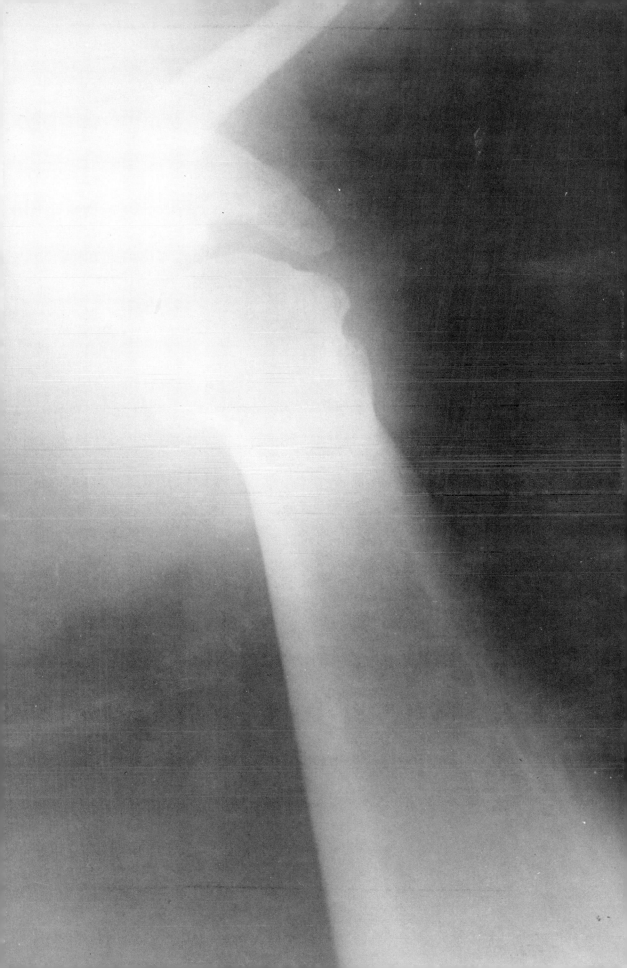

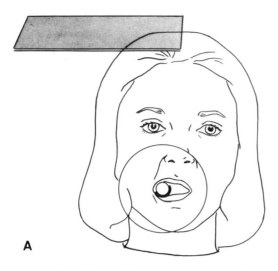

A

Fig. 7-10A. Spot of the zygomatic arch. The Panoramix tube is positioned in the mouth, and a 14 x 17 inch cassette is placed on the top of the head.

REFERENCES

1. A Study of Head and Neck Exposures From Panoramix Roentgenography. U.S. Department of Health, Education, and Welfare, Public Health Service, No. BRH/DEP 70-5. Washington, D.C., U.S. Government Printing Office, March 1970.
2. Cullinan, J. E.: Westinghouse Panoramix Model MO-200: Technique/Application Manual. Baltimore, Westinghouse Electric Corporation, X-Ray Division, n.d.
3. Isard, H. J., Brotman, M. S., and Cullinan, J. E.: Panoramix roentgenography. X-Ray Bulletin AGFA-Geveart, No. 7, pp. 3–17, 1967.
4. Schultz, E. H., Jr., and Wood, E. H.: Mobile x-ray equipment. *In* Scott, W. G. (ed.): Planning Guide for Radiologic Installations. ed. 2, pp. 103–106. Baltimore, Williams & Wilkins, 1966.
5. Trew, D. J.: The "cordless" mobile x-ray machine. Radiography, *XXXVI*(427):159–164, July 1970.
6. Trinks, W.: The Panoramix 360° method. Cathode Press, *22*(3):28–29, 1965.
7. Updegrave, W. J.: Panoramix Dental Radiography. Dental Radiography and Photography, Rochester, N.Y., Eastman Kodak Company, 1963.
8. Vennes, C. H., and Watson, J. C.: Patient Care and Special Procedures in X-Ray Technology. pp. 94–95, St. Louis, C. V. Mosby, 1964.
9. Young, B. R., and Scanlon, R. L.: New explosion-proof and shock-proof mobile roentgenographic equipment for the operating room. Amer. J. Roentgen., *71*:873–877, 1959.

Fig. 7-10B. Enlargement of the submental vertex view of the malar bone, zygomatic arch, and adjacent structures. Several fractures can be seen.

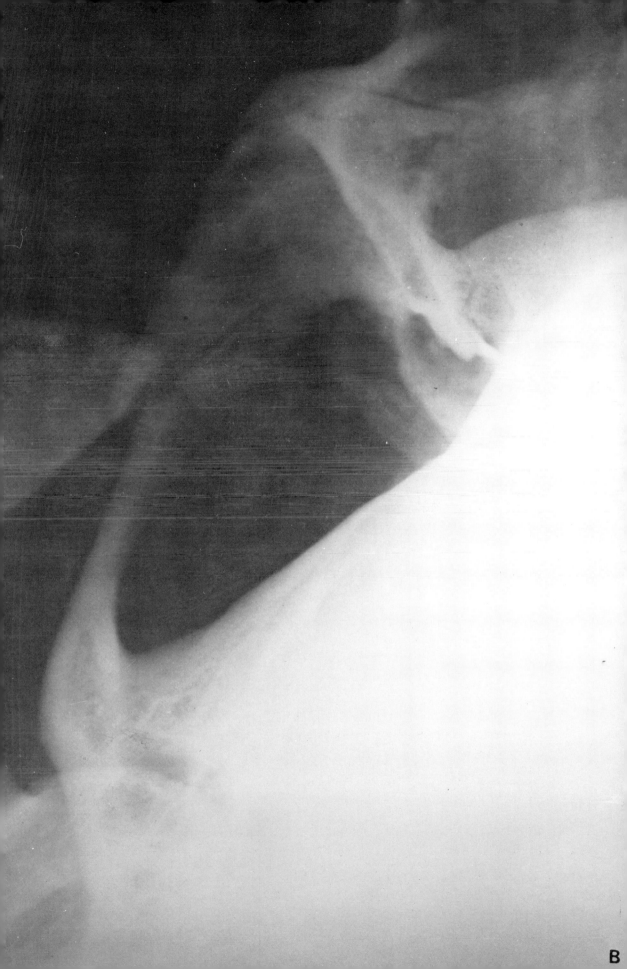

B

8. Body Section Radiography

DEFINITIONS, PURPOSE, AND SCOPE

Body section radiography is a complex technic which is used to locate a predetermined level or plane within the body.[1] Superimposed radiographic shadows are erased or blurred from the finished radiograph or body section film. "Tomography" is the term selected by the International Commission of Radiological Units and Standards as the generic term to designate all systems of body section radiography.[18] The term "tomography," derived from the Greek word *tomos,* meaning a *cut* or a *section,* was coined by Grossman in 1935.[9] Bocage, the originator of tomography, was the first to describe it correctly in his application for a patent in 1921.[2] There are many interchangeable terms for body section radiography; some of these are "planigraphy" (from *planum,* meaning *plane*), "stratigraphy" (from *stratum,* meaning *layer*), and "laminography" (from *lamina,* also meaning *layer*).

"Body section radiography" is a generally accepted term for the use of the contrary movement of an x-ray source and of film to record a section with relative clarity that otherwise might be obscured by an object in a layer above or below the selected plane.[21] For all types of body section radiography, regardless of the type of tube movement, the x-ray tube and the film move synchronously opposite to each other during the exposure. They are connected by some type of mechanical device to an adjustable fulcrum, which permits the selection of the desired level of the body to be radiographed (Fig. 8-1). Selecting and visualizing a plane within the body at a predetermined level is made possible by the fact that all planes above and below the fulcrum level are not recorded on the film in sharp focus but rather as a blur, or with some type of streaking pattern, depending on the tube-film motion used.

Often a conventional radiograph will exhibit more information than is required for interpretation. For example, if a lesion of the lung is noted in the right upper lung field, extraneous yet normal radiographic densities such as ribs, vessel markings, etc., are superimposed upon the area of clinical interest. With body section radiography these shadows are blurred out, whereas the lesion is kept in relatively sharp focus for evaluation. Superimposed images are removed by body section radiography, for as the tube and film move, there is a widening or broadening of the margins of the images above and below the fulcrum point, together with a marked reduction in contrast of the overlying images.

Ring, quoting Dr. Irving Kane in his book *Sectional Radiography of the Chest* (Springer, New York, 1953) describes body sectional radiography as a method of "bloodless dissection."[17] Dr. Kane states that this method yields information obtained by no other procedure, and that sectional radiography separates shadows and affords a picture which resembles that seen only by a pathologist.

As tomographic units become available with elaborate obscuring motions, body section radiography approaches Dr. Kane's description in the truest sense of the word. The thinner and thinner cutting of sections makes

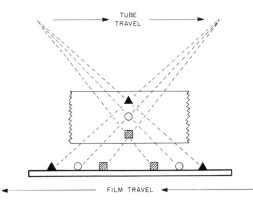

Fig. 8-1. Basic concept of linear tomography. Both the x-ray tube and the radiographic film must be in motion throughout the entire exposure. The tube moves in one direction, and the film moves in the opposite direction about a pivot point or fulcrum. Anything not in the plane determined by the level of the fulcrum is displaced relative to the plane and is deliberately obscured by a blurring motion. The circle is in plane, whereas the triangle and the square are displaced from their respective positions.

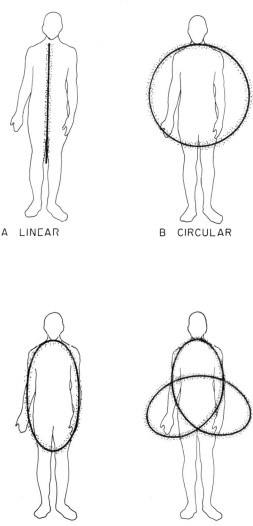

Fig. 8-2. Tube-film trajectory. Some tube-film travel patterns are quite complex. Basic tube-film paths include:

(A) *Linear.* This is the most common unit in use today. Tube and film move in a straight line.

(B) *Circular.* The circular motion significantly increases the *exposure angle*, as compared with the typical linear motion.

(C) *Elliptical.* This is an elongated motion that results in a generous *exposure angle*.

(D) *Hypocycloidal.* This is an asymmetrical cloverleaf trajectory, which is 5 times greater than that of the conventional linear route. Extremely thin sections are possible with a unit using the hypocycloidal movement.

almost possible the use of the term "radiographic biopsy." If a complex obscuring motion could be combined with direct-image enlargement technic, the term "radiographic biopsy" would be even more descriptive.[6, 16]

Tube-Film Trajectory

Although the blurring of superimposed anatomic details is accomplished by the moving of the x-ray tube and the x-ray film in opposite directions to each other throughout the entire length of the x-ray exposure, different types of x-ray equipment have different tube-film motions. Some of these motions include linear, elliptical, circular, and hypocycloidal (Fig. 8-2). As the motion of the tube and film become more complex, both the blurring capability and the cost of the unit increase. The most common body section unit in use is not a complex piece of equipment but an attachment to an existing table. Its tube-film motion is linear.

Multidirectional units that utilize hypocycloidal, circular, or elliptical motions will be reviewed in this chapter, but the principles of the linear unit will be stressed. Definite advantages and limitations of the linear systems, as well as representative radiographs, will be shown. A brief review of tube-film motion follows.

1. **Linear Movement.** A linear tomographic attachment can be added to almost any ordinary radiographic unit at a very reasonable cost. As with all methods of tomography, the film must move in exactly the same path but opposite to the motion of the x-ray tube. Many of these units can be energized from a small arc (10 degrees) to a maximum exposure angle of 50 degrees. One of the difficulties encountered with linear tomography is the restriction on the length of tube travel. Striations or linear streakings occur with this type of system. Since the focal spot and film move in straight parallel lines, shadows of rod-like objects parallel to the direction of the tube motion, even though out of focus,

may be reproduced on the finished radiograph. Maintaining mechanical stability and vibration-free operation with a linear unit is difficult if the unit is attached to an x-ray tube.

2. **Circular Movement.** A distinct advantage of circular movement over the linear movement is that the total exposure angle can be increased, for the length of tube travel is increased by means of the circular tube travel path. Occasionally, the circular movement may give rise to misinterpretation, as a small circular object may appear on the film as a large circular shadow.

3. **Pluridirectional Motions.** Pluridirectional obscuring motions provide a significantly better blur effect than linear tomographic movements. These units are generally self-contained, are not attachments to existing pieces of x-ray equipment, and require at least a large portion of a standard radiographic room. Some units utilize the

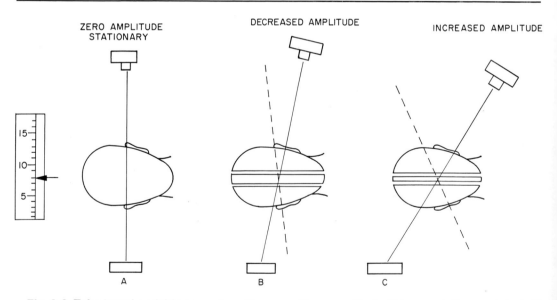

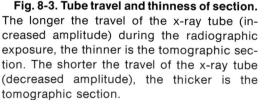
Fig. 8-3. Tube travel and thinness of section. The longer the travel of the x-ray tube (increased amplitude) during the radiographic exposure, the thinner is the tomographic section. The shorter the travel of the x-ray tube (decreased amplitude), the thicker is the tomographic section.

Zero amplitude (A) or a stationary tube will result in a lateral projection of the skull. Decreased amplitude (B), 10 to 20 degrees, results in a thicker section of the skull. Increased amplitude (C), 40 degrees or more, results in an extremely thin section of the skull.

linear, circular, elliptical, or hypocycloidal motion, or combinations of these methods (Fig. 8-2).

The most elaborate pluridirectional tomographic unit available is the Polytome (Massiot Philips), which was introduced in France in 1951 by Raymond Sans and Jean Porcher.[21] It offers a choice of 4 types of obscuring motions—linear, circular, elliptical, and hypocycloidal. The hypocycloid is defined as the plane curve generated by a point on a circle rolling inside a larger circle.[19] The hypocycloidal movement is asymmetrical and is 5 times greater than the conventional linear movement.

A definite disadvantage of the hypocycloidal motion is the long time required for adequate exposure of the x-ray film. A typical exposure length would be 6 seconds. It is helpful in using the hypocycloidal movement to ask the patient to keep his eyes closed during the lengthy exposure, so that he will not tend to follow the complex movement of the x-ray tube.[21]

Terminology

New terminology is required to understand body section radiography. Some of the terms used include:

Objective plane is the plane whose shadow is stationary as related to the film. The finished radiograph of this plane is called a "tomographic section."

Focus-film distance is a familiar term. It is used in body section radiography as well as in conventional radiography to describe the distance between the x-ray tube focus and the x-ray film.

Focus-plane distance is described as the distance between the x-ray tube focus and the objective plane.

Plane-film distance is the distance between the objective plane and the radiographic film.

Thickness of Section

Although the layer or section to be radiographed is determined by adjusting the fulcrum, the thickness of the section is controlled primarily by *amplitude*. "Amplitude" is

defined as the distance that the x-ray tube travels during the radiographic exposure. *The longer the travel of the x-ray tube, the thinner is the section; the shorter the travel of the radiographic tube, the thicker is the section* (Fig. 8-3). Another factor that controls the thickness of the section is the variation of *focus-film distance* (Fig. 8-4). Variation in focus-film distance is rarely used, but in principle the closer the x-ray tube to the patient, the thinner is the section. Although amplitude is used to describe the length of the tube movement during the exposure, the angle through which the x-ray beam moves during the exposure is known as the *exposure angle* (Fig. 8-4). Exposure angle must be increased when the thinnest possible

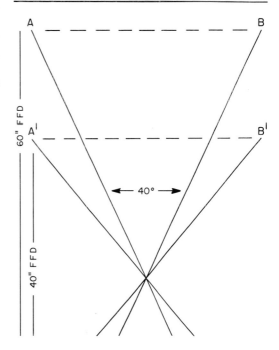

Fig. 8-4. Variation in focus-film distance and thinness of section. When a fixed FFD—for example, 60 inches—is used, a fixed exposure angle of 40 degrees will result as the tube travels from A to B.

When the FFD is decreased to 40 inches, the exposure angle is increased if the same length of tube travel is maintained from A' to B'. Since the exposure angle has been increased with the shorter FFD, thinner tomographic sections will result.

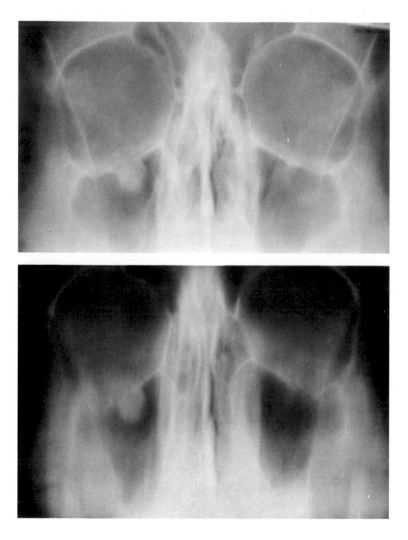

Fig. 8-5. Zonography versus thin section tomography. Sectional films were made of the facial bones and orbits. (*Top*) A 10 degree amplitude (zonogram), with excellent visualization of the orbits. (*Bottom*) A 40 degree section was made, and a thinner section is the result.

Zonography is quite adequate for examining the orbits and facial bones as compared with thinner section tomography. An added advantage of zonography is the fact that conventional radiographic factors can be used at 10 degrees, and good radiographic contrast can be maintained.

cut is necessary. If the exposure angle is reduced to zero degrees (zero amplitude), the result is a normal radiograph (Fig. 8-3). Littleton recommends the monitoring of the performance of the tomographic device during its useful life. An appropriate phantom has been designed for this purpose.[12]

Why Tomography?

Two basic reasons for resorting to tomography are *analysis* and *disengagement* of the radiographic image.[25] Analysis is the objective when tomography is used to disassociate and display separately the anatomic details of a complex structure. Disengagement is the goal when tomography is used as a means to erase image details of no interest belonging to structures in the path of the beam so that the structure of interest can be seen better (Fig. 8-5). In using tomography with analysis in mind, a multitude of thin cuts are necessary. An excellent example of analysis is the study of the petrous bone for middle and inner ear changes (Fig. 8-6). Tomographs made in this area or any other area of the body with a 40 or 50 degree exposure angle result in cuts within 1 mm thickness. An example of disengagement would be tomography of the kidney at the nephrographic stage. On occasion, films made utilizing a zonographic technic (a 10 degree or smaller exposure angle) are to quite helpful during intravenous urography. Tomographic cuts of this type can eliminate superimposed gaseous shadows if a patient has not been fully prepared for a radiographic study[24] (Fig. 8-7).

A section utilizing a 10 degree exposure angle results in a layer of approximately 1 cm. If the exposure angle should be decreased to 5 degrees, the layer visualized will be approximately 2 to 3 cm thick.

There is a tendency to generalize or to group anatomic areas with specific exposure angles. Many feel that all cuts taken within the cranial vault or of the facial bones should be relatively thin. A thick cut or zonogram of the orbits can be quite informative, however, because of their complex shape. Increased angle tomography of the order of a 40 degree cut will demonstrate only a limited portion of the floor of the orbit (Fig. 8-5). Regardless of the type of study desired, whether the primary purpose be analysis or disengagement, the direction of the obscuring movements of the unit must be considered, since portions of the body under study parallel to the direction of the tube-film motion can never be completely erased. "Parasitical"

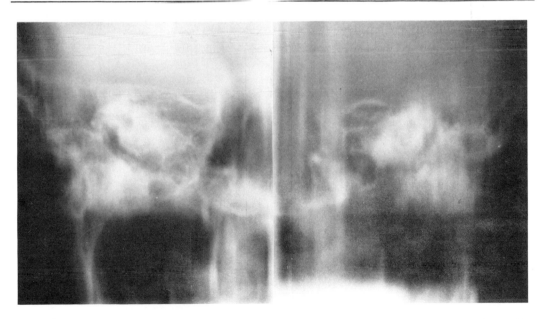

Fig. 8-6. Petrous ridge tomography. Two separate films were made of the petrous ridges. in the film on the left, which was made with a 10 degree exposure angle, almost the entire bony ridge is seen, completely disengaged from the skull.

In the study on the right, which was made with a 50 degree exposure angle, an extremely thin section is available for interpretation. The delicate structures housed within the petrous ridge are available for analysis.

Zonography (10 degree exposure angle) results in a *gross* specimen for evaluation, whereas increased angle tomography (40 degrees or more) results in a *radiographic biopsy.*

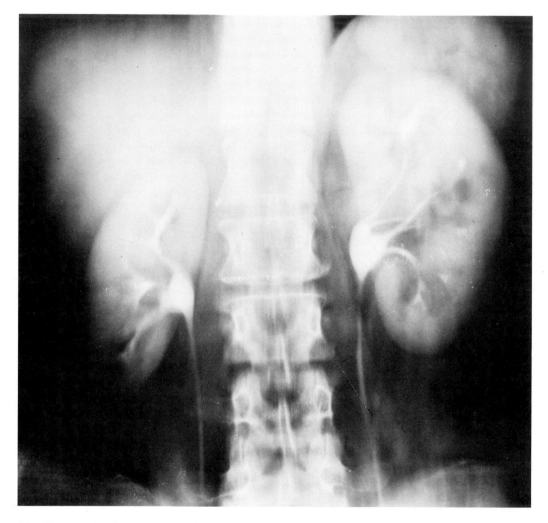

Fig. 8-7. Nephrotomogram. A thick section (20 degree exposure angle) made during a drip infusion pyelogram. Narrow angle tomography (5 to 10 degrees) can be quite helpful in examination of the urinary tract. Complete erasure of overlying gaseous or fecal shadows is possible with this technic.

images caused by residual streaking hamper the sharpness of the finished image. The proper selection of a scanning movement is helpful when maximum obliteration of these residual images is necessary.

To be completely accurate, a body section study must often encompass the entire lesion or structure under evaluation. Unless an adequate number of sections are made, a questionable finding on a conventional film will be ignored on the basis that the tomographic study is seemingly adequate. The lumbar vertebrae and biliary ductal system are prime examples of this technical difficulty (Figs. 8-8 and 8-9A, B, and C).

ZONOGRAPHY

Zonography is used to "remove" (disengage) a lesion from its surroundings, whereas tomography enables the beholder to "see" (provides analysis) into that lesion.[20] In effect, is a gross radiographic specimen desired or a radiographic biopsy? Zonography may be used to localize a structure or lesion before definitive (analysis) tomography is attempted.

With a small exposure angle (5 to 10 degrees) a thicker layer, though relatively sharp, approximately 2 cm thick, is literally lifted out of the body.[25] Zonography can be quite helpful in examining critically ill patients for fractures of the mandible or facial bones.[26] Not only does zonography save considerable time, but it provides much diagnostic information about thicker sections of anatomy (Fig. 8-5). This technic requires a smaller quantity of film and a reduction in the amount of radiation exposure, and at the same time it supplies an improvement in radiographic contrast. With the use of zonographic technics there is a definite decrease in image purity but a significant increase in contrast.[22]

It is easy to perform linear zonography with a 5 or 10 degree exposure angle. A minor adjustment in the electrical mechanism that controls the x-ray exposure angle permits the interested technologist to perform zonography. Five or 10 degree angle linear-movement cuts utilize exposure times comparable in length to those of conventional studies. Zonograms resemble conventional radiographs. Mechanical stability with zonography is excellent as a result of the shortened tube-film travel. Because of the reduced angulation of the tube in relation to the area under study, distortion is reduced. Tube-travel time is realtively short, and therefore the exposure time can be correspondingly shortened. With linear zonography there is a general reduction in striations or at least in the length of the striations[15] (Fig. 8-5). With some of the more elaborate pluridirectional devices, zonography can be performed with a small angle circular movement.

As an increased exposure angle is used the exposure factors are increased; for the body thickness varies during the longer tube-film run. For example, with a 40 degree cut of an AP abdomen (approximately 18 cm) at the start and end of the exposure the central ray enters the abdomen at an angle of ap-

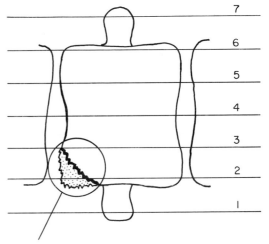

DESTROYED BONE

Fig. 8-8. Necessity of multiple sections. A 7 cm segment of a lumbar vertebra is used to stress the need for multiple section tomography. Often a body section study must encompass an entire lesion or structure to be diagnostically accurate, for unless an adequate number of sectional films are made, a questionable finding on a conventional film will be ignored because of an apparently normal tomographic study.

The osseous pathology in the lumbar spine would be demonstrated on section 2 and might be seen on section 3. If films were taken at levels 4, 5, and 6, they would demonstrate normal bone and could be misleading. *Be certain that a complete set of films is made during the initial tomographic study.*

proximately 23 cm. As the central ray approaches the midpoint where the minimal amount of blurring occurs, the patient thickness becomes normal, and there is normal patient-beam attenuation. To compound further this variable thickness difficulty, the focus-film distance increases as the radiographic tube moves away from the center position.

SOME USES FOR BODY SECTION RADIOGRAPHY

For many years tomography was restricted to examination of the chest. Recently, the use

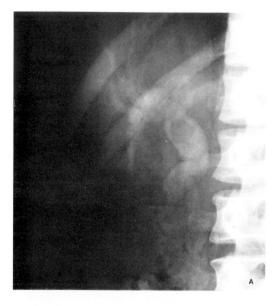

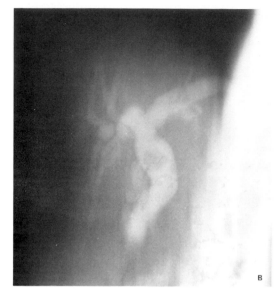

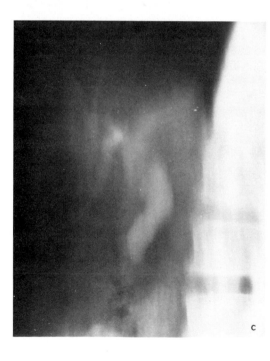

Fig. 8-9A. Intravenous cholangiography, conventional film. This radiograph was made 40 minutes after the injection of an opaque contrast agent. Visualization of almost the entire biliary system is excellent. A large radiolucent defect can be seen in the upper portion of the common duct.

Fig. 8-9B. A linear tomogram taken within minutes after the film shown in 8-9A. The contrast material has been definitely enhanced by the use of body section technic. The radiolucent defect in the upper segment of the common duct is quite obvious. Note that the lower end of the biliary ductal system is now unsharp in this tomographic cut, and it would be impossible to confirm or deny the absence of further radiolucent defects in that area.

Fig. 8-9C. An additional tomographic cut 1 cm away from 8-9B. Note the smooth tapered end of the ampulla of Vater in excellent focus. The entire ductal system has now been examined by means of multiple sections, but the large lucent defect seen in the previous radiographs is now almost completely obliterated by a fulcrum adjustment of only 1 cm.

of tomographic equipment for such studies as intravenous urography or cholangiography has become quite commonplace. Many technologists still feel that body section radiography is a special procedure and treat it as such. Yet there are many uses for tomography which would help the practicing technologist to supplement a series of radiographic films for easier interpretation by a radiologist. Some of these uses include:

1. **Intravenous Urography.** When gaseous shadows obscure the kidneys or any segment of the urinary system, a tomogram can help to overcome inadequate preparation of the patient. Since the purpose of the study is to evaluate the entire kidney rather than a thin section of the kidney, zonography can be helpful (Fig. 8-7).

2. **Intravenous Cholangiography.** The use of a tomographic device with intravenous cholangiographic studies is of the utmost importance. Often the biliary ductal system will exhibit very weak contrast after the injection of an opaque iodine substance. Yet biliary ducts frequently will be demonstrated on a tomographic section even if they have not been seen on the conventional films because of a low iodine concentration (Fig. 8-9).

3. **Orthopedic Procedures.** Body section radiography has proved to be of value to the orthopedist.[7] Some uses include:

The Ability to Remove a Plaster Cast from a Patient Radiographically Rather Than Physically. To remove a body cast from a patient who has had a serious injury such as a fractured hip is time consuming, and can be harmful to the patient if the stability of the fracture is in question. The cast can actually be "removed," that is, obliterated or blurred by tomography, and the fracture site evaluated layer by layer.

Demonstration of Fractures of the Metacarpal Areas after the Application of a Cast. A casted hand or wrist can be placed in a true AP or lateral projection, and a tomographic cut made that completely obliterates extraneous details. When examining the metacarpal area, we make a steep oblique projection instead of a true lateral projection. With

tomography, however, a true lateral projection can be made whether the hand is covered with plaster, is bandaged, or is in a splint. This technic can be used also for AP and lateral views of the fingers through a plaster cast (Fig. 8-10).

Radiography of the Patella. Radiography of an injured patella can be difficult if a patient is unable to flex his knee. Tomographic films can be made of the knee prior to or after casting for the position of fragments.

SOME BASIC TECHNICAL CONSIDERATIONS

Certain basic technical considerations are common to all methods of tomography. Emphasis will be placed on linear tomographic technics. Major concerns must include:

1. **Patient Positioning.** Body section radiography is often limited to the AP or PA position. On occasion, when a specific lesion in a specific area is better demonstrated in a conventional oblique or lateral position, the tomographic study should be made in that position. Examine the original radiographs, and if the lesion in question is not visible in the standard frontal projection but is visible in the lateral projection, it may be of some value to tomograph the patient in the lateral projection. To section the chest in the AP position, particularly in the lower lung field near the level of the diaphragm, is sometimes difficult because of the need for increased technical factors. It is easier to section the patient in the lateral projection. The lateral projection can be very helpful in trying to demonstrate lesions of the chest just posterior to the sternum.

2. **Level Determination.** Determining the proper level of cuts from a routine radiograph is difficult. Level determination should be made from either the PA or the AP radiograph that initially demonstrated the lesion, and a scout tomographic cut should be made at what would seem to be the proper level. When the initial cut has been made, another film should be exposed either 2 cm in front of or behind the initial cut. Both scouts are

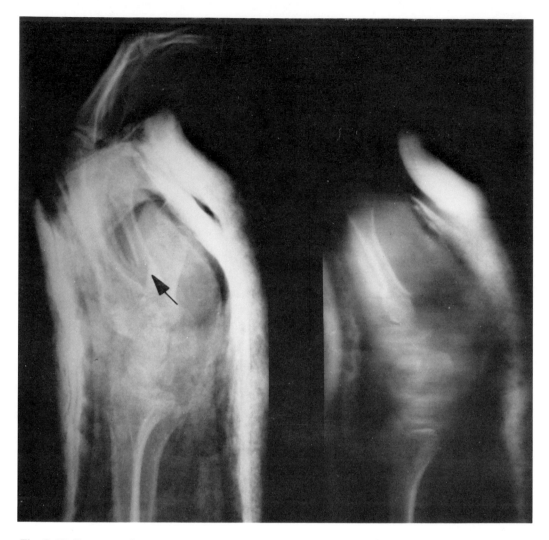

Fig. 8-10. Tomography of a casted extremity. Shown are body section films of the hand made at a 20 degree exposure angle in the true lateral projection. The remaining bones of the hand as well as a plaster cast are completely obscured by tomography. A fracture of the 5th metacarpal is demonstrated quite clearly in the lateral position.

The use of tomography to "remove" a plaster cast is particularly helpful when a patient with a fracture of the femur, hip, humerus, etc., is being examined, and it is important to maintain stability of the fracture fragments. It is easier to section a body part while it is casted than to remove a patient physically from a cast and jeopardize healing and/or alignment.

viewed immediately, and level determination is made. Two cuts are taken, for even if the 1st section demonstrates that the lesion is in, or almost in, the plane, one cannot be certain whether the tomographic cut was made through the center, anterior to, or posterior to the lesion. Evaluation of the 2nd scout film reveals that the lesion is in sharp focus, in

poor focus, or not in focus at all. The direction for continuing the tomographic study can then be determined (Fig. 8-11). If one were to make the 1st cut posterior or anterior to a lesion, it would be virtually impossible to determine in which direction to reradiograph without an additional cut. The 2nd cut serves another purpose, which will be explained in the next topic, Factor Determination.

Although chest radiographs are taken at a 72 inch FFD with the patient in the PA erect position, most tomograms of the chest are made in the supine AP position with a 40 inch FFD. To measure accurately the location of a lesion on a conventional chest radiograph for tomographic purposes is a complicated task.

Technologists may be called upon to tomograph a lung lesion and to determine the exact fulcrum level for the study. If preliminary radiographs are not available, it is absurd to attempt a tomographic study.

If the lesion is seen in both the frontal and the lateral projection, level determination is relatively easy. Divide the lateral projection into 4 quarters to determine the location of the lesion (perhaps the lesion is slightly anterior to the 4th section of the chest) (Fig. 8-12). The patient is placed in the AP supine projection, and the exact area to be measured is determined by marking first the PA radio-

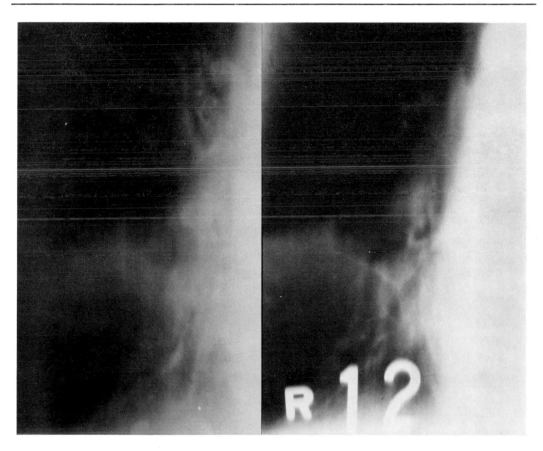

Fig. 8-11. Technic and level determination for tomography. Two scout films can often expedite a tomographic study. The area to be evaluated is centered to the film, and the 1st scout film is made at a predetermined level with specific technical factors. A 2nd scout film is then taken, with a 10 kVp increase in technic, either 2 cm above or below the level of the 1st survey section.

The film on the left was made at 55 kVp at a 10 cm level from the tabletop, and the film on the right was made at 65 kVp at a 12 cm level from the tabletop. A radiopaque wedge-shaped area is sharply defined in the parahilar region on the 12 cm film.

Level determination can now be more accurately established. Sections can be made on either side of the 12 cm level. Technical factors can be adjusted after evaluating both scout films.

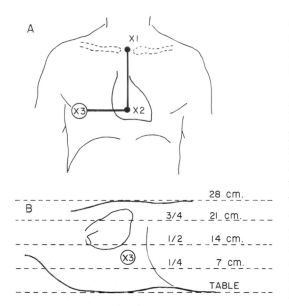

Fig. 8-12. Simple tomographic localization technic. The lesion in question (X3) is seen in the lateral aspect of the right lung field on the initial PA chest radiograph. A crayon mark is made on the film (X1) at the level of the sternoclavicular joints. A second mark (X2) is made in the midline at the level of the suspected lesion (X3). These measurements are transferred to the chest of the patient. The chest is then measured at X3, and a 28 cm measurement is obtained. The lesion (X3) is localized on the lateral radiograph, which has been divided into 4 quarters of 7 cm each. The lesion (X3) is localized in the 2nd quarter between 7 and 14 cm. Scout cuts are made at 9 and 11 cm.

graph with a wax crayon at the level of the sternoclavicular joints (X1). A measurement is taken at the level of the lesion, and a 2nd mark is made (X2) directly beneath the sternoclavicular notch (X1). An additional measurement is then taken from the lesion (X3) to the 2nd marking (X2). A crayon mark is now made on the patient at the sternoclavicular notch, and the film markings are transferred to the skin of the patient. In practice it is easier to use the thumb and forefinger to make these measurements on the film from the notch to the level of the lesion, and from the level of the lesion to the

actual lesion, and to transfer these measurements to the skin of the patient. *The chest is then measured through the exact area (X3), which is marked with the crayon.*

This particular patient measured 28 cm (at X3) in the AP supine projection (Fig. 8-12). Hence each quarter segment of the lateral projection measures 7 cm. Since the lesion is slightly anterior to the 1st quarter of the lateral projection (theoretically 7 cm from the tabletop), the initial section will be 9 cm from tabletop. The 9 cm section is made with the central ray (zero amplitude) entering the predetermined mark (X3) on the patient. A 2nd radiograph can be made at the 7 or 11 cm level with an increase in technical factors. Both scouts are immediately checked, the level determination is finalized, and minor technical adjustments are made.

A simple method for determining level is the utilization of a Polaroid cassette or similar rapid-process cassette. The film is divided in half, and 2 cuts are made on a single sheet of Polaroid film. The film is processed, and level determination is immediately made. Unfortunately, factor determination cannot be easily made with the Polaroid study because of its increased film speed. With the advent of 90 second processing, the split film technic using conventional screen and film becomes a reality.

Making 2 exposures to determine level and technic is better than making a series of overexposed or underexposed radiographs. Proper film density, exact level determination, and heat unit production must be individually considered prior to a tomographic study.

When *analysis* is requested, multiple thin cuts, relatively close together ($\frac{1}{4}$ to $\frac{1}{2}$ cm), can be made after an exact level has been established. If *disengagement* (small-angle tomography) is attempted, level determination is significantly easier to achieve.

3. **Factor Determination.** How does one determine the factor combinations of kVp, mA, and time required for body section radiography? A simple method is as follows: Using the dual scout technic (see the pre-

ceding topic, Level Determination), expose the 1st film at a predetermined technic. The 2nd scout, varying 2 cm from the 1st cut, could be increased approximately 10 kVp.

It is often more acceptable to increase kVp rather than mAs for several reasons: First, there is lower radiation dosage to the patient. Second, significantly fewer heat units are generated in the x-ray tube with the use of higher kVp rather than higher mAs values. Third, most body section radiographs utilize a well collimated x-ray beam; therefore an increase in kVp is generally not technically harmful. As a result of the 2 cm level variation and the 10 kVp increase, 2 determinations can be made: first, the proper level, or at least the proper direction, in which to obtain the proper level, and, second, the exact or at least adequate radiographic factors from which appropriate technical adjustments can be made. From the 2 scout radiographs 2 major technical decisions have been made: *level determination* and *factor determination* (Fig. 8-11).

How does one formulate starting radiographic factors for body section studies? In general, the average body section film requires approximately a 50 per cent increase in the mAs value over the conventional film when the original kilovoltage value is retained. For instance, if an intravenous cholangiographic film required 80 kVp at 100 mAs, 150 mAs with the same 80 kVp value would be required for the tomographic cut (50 mA at 3 seconds would be adequate). If a lateral film of the tempromandibular joint required 50 mAs at 75 kVp, an increase to 75 mAs at the same kilovoltage level (25 mA at 3 seconds) would be required for body section radiography. Kilovoltage increases shorten exposure time, but unfortunately most body section devices require long scanning times. Very few tomographic units can travel over a 40 degree arc in as quick a time as a half second.

When conventional technics are converted to tomographic values, certain requirements must be met. It is assumed that when these adjustments are made, (1) there is no change in the field size of the beam used; (2) the same type of collimator or its equivalent is being used; (3) there is no increase or significant decrease in the focus-film distance; (4) a grid of the same ratio is being used; and (5) the screen or film combination is of the same speed as that used for the conventional radiograph. Technical adjustments must be made if there is any variation in the previously mentioned requirements.

Another consideration in determining body section factors is the portion of the body that is absorbing primary radiation. If the technologist were to tomograph the soft-tissue structures of the neck in the supine position, using an extended exposure angle, the central ray could pass through the apical portion of the chest, the cervical spine, the facial and the occipital portion of the skull, and would require a significant increase in exposure over conventional cervical spine factors. If the technologist were to radiograph the same patient so that the x-ray tube would move at a right angle to the neck rather than parallel to the neck, only the cervical spine would be penetrated instead of the skull, cervical spine, and chest. Reduced tomographic exposure factors would be needed with this second technic. Technical factors when using 5 or 10 degree zonographic cuts approximate conventional factors.

Some Technical Difficulties Encountered With Tomography

Since many tomographic units are not permanent installations, providing a rapid transition from radiography to tomography is difficult, especially in view of the amount of time required to attach a temporary unit to a radiographic table. A busy department should not be without a permanent tomographic installation. When an attachment is used for tomography, it may not be stable while it is in motion. Tube centering as well as accuracy of alignment can also be a problem. When small field-size tomography is performed, careful alignment of the light beam collimator must be assured.

Intensifying Screen-Film Combinations

Conventional medium-speed intensifying screens are satisfactory for tomographic studies. There is no advantage in the use of slow or fine-grain intensifying screens, for they do not usually improve the image.[10] The sharpness demonstrated on the body section radiograph (due to the blurring movement) is significantly greater than the sharpness potential afforded by the typical intensifying screen. Reducing exposure by the use of a faster film is better than using fast-speed intensifying screens.[20] Radiographic definition can be affected by parallax, which can be overcome by using a single-coated film and a single screen; these minimize the unsharpness associated with angulation of the beam and the duplitized film-screen combinations.

Radiographic Contrast

The loss of contrast with conventional tomographic technics is appreciable. With an increase in the blurring movement of the tube and film, a further diminution of contrast occurs. It is a common belief that the kilovoltage range should be kept at a low to moderate level for maximum contrast during body section radiography. If moderate kilovoltage values are used with a corresponding increase in milliampere-second factors, films exhibiting severe contrast are produced because of the improved scattered radiation clean-up capability of available grids and light beam collimators.

This author feels that an increase in kilovoltage is not objectionable with today's high-ratio grids and primary ray collimators. Seeman feels that technologists should be made to use lower kilovoltage values—even as low as possible—and stresses the limiting of the x-ray beam to its minimum useful port size. He feels that much of the radiation passing through the patient serves only to blur out unwanted images but does not contribute directly to the formation of the useful image, although it does expose the film. Physically this exposure is like that of scattered radiation lowering subject contrast.[19]

Technical Adjustments. Adjustments must be made when the technologist is converting radiographic factors from conventional to tomographic studies. For example, most chest radiographs are taken at 72 inches FFD. Since a 40 inch FFD is used for most body section films, it is first necessary to lower the mAs value to approximately one fourth of its original value. Second, many departments use high kVp chest technics (125 kVp), and yet moderate kilovoltage levels are stressed for body section radiography. Third, the restriction of the primary ray to small ports, as opposed to the original 14" × 17" beam, theoretically necessitates an increase in exposure values. Fourth, if an exposure angle of 40 degrees were to be used for thin sections, there would be a change in tissue thickness in relation to the beam, necessitating a corresponding increase in the factors required.

It is difficult to make technical adjustments with typical milliampere stations. Generally, the minimum mA value on most radiographic units is 50 mA. If a linear tomographic unit is required to operate from 2 to 4 seconds to produce a 40 degree cut, the mAs values for chest radiography would be 50 mA × 2 seconds (100 mAs), or as much as 50 mA × 4 seconds (200 mAs). Even at moderate kilovoltage levels these factors would be excessive for radiography of the lung. The only decision left, without changing film or screens, is to lower kilovoltage. If lung tomography is to be successfully accomplished with a slow-moving linear tomogram, it is not possible to use such factors as 50 mAs at 60 kVp with a 2 to 4 second exposure time. The 50 mA station could be changed to 25 mA for a 2 second pass (in the 60 kVp range), factors which would still produce a radiograph of extreme contrast. The utilization of a 12 or 15 mA station with a 2 second pass (in the 70 kVp range) would produce more nearly acceptable tomograms of the lung, or a 5 mA station with a 3 second run in a similar kilovoltage range would be equally appropriate. When the entire lung field is being scanned (14" × 17") for visualization of the vasular patterns of the lung or pulmo-

Table 8-1. Comparison of a Typical Tomographic Study and a Serial Angiogram

LATERAL SKULL CEREBRAL ARTERIOGRAM (12 FILMS)	LATERAL SKULL TOMOGRAPHY (5 SECTIONS)
400 mA, ½₀ second, 80 kVp (48" FFD) = 1,600 heat units per exposure	50 mA, 2 seconds, 65 kVp (40" FFD) = 6,500 heat units per exposure
1,600 heat units times 12 exposures (2 per second for 6 seconds) = 19,200 heat units	6,500 heat units times 5 exposures = 32,500 heat units

COMPARISON	
Lateral skull tomogram (5 exposures)	32,500 heat units
Lateral skull angio (12 exposures)	19,200 heat units
Excess H.U. of tomograms over the angios	13,300 heat units
Lateral skull tomogram (12 exposures)	78,000 heat units
Lateral angio (12 exposures)	19,200 heat units
Excess H.U. of tomograms over the angios	58,800 heat units

nary metastasis, a 5 mA or 10 mA station is a definite advantage. This large port with its corresponding increase in scattered radiation demands a low mA value if an extended scan technic is used. Some tomographic attachments can make a half to 1 second pass, but frequently a slower tomographic pass is used to ensure the mechanical stability of the unit.

Body Section Heat Unit Considerations

It is a common mistake to expose 5 or 10 body section films without concern for the anode heat capacity of the radiographic tube. Tomography can cause more damage to an x-ray tube than a rapid serial angiographic study in which fast-speed intensifying screens in combination with fast film are utilized to shorten exposure time. See Table 8-1.

As demonstrated in Table 8-1, if a 12 cut tomographic study were to be attempted, there would be an increase of 58,800 heat units over the lateral cerebral angiographic technic. In the more typical 5 section tomographic study of the skull, there would still be an increase of 13,300 heat units over the angiographic technic (12 serial exposures).

An interesting heat unit problem occurs with a very popular tomographic unit—not a tomographic attachment but a tomographic unit with a fixed 57 inch FFD. If one were to take 5 tomographic cuts of an average patient in the lateral position to demonstrate

the lumbar spine, the following factors are required for individual cuts: 200 mA, 4½ seconds at 90 kVp, resulting in a heat unit factor of 81,000 heat units per exposure. The total, then, of 81,000 heat units per exposure times 5 exposures is 405,000 heat units. The maximum anode heat capacity of one of the largest tubes available that can be mounted on this unit is 140,000 heat units. Five cuts of the lateral lumbar spine could destroy this tube. *The total heat units involved in a study must be determined prior to starting a tomographic examination.* If 3 sections of an intravenous cholangiogram are made with factors of 100 mA at 2½ seconds in the 80 kVp range, 20,000 heat units per exposure would be generated. Three exposures at 20,000 heat units each would result in a total of 60,000 heat units. Almost any tube could tolerate these factors.

Heat units must be totaled prior to the beginning of a study, and the technologist must pace the exposures during the examination. Repetitious cuts can be quite boring, and technologists will work together to complete a study, not allowing adequate anode cooling time. The most common error encountered when heat units have been determined in advance is the taking of additional films without regard for further heat unit generation. For example, (1) a technologist can forget to use the proper lead number on a film to indicate level, and decide that it is important to re-expose the patient with the proper marker in place; (2) the proper level can be chosen, and the wrong level marked on the film; (3) improper factors can be used; (4) the patient can move or breathe, so that an additional cut becomes necessary. Even though the technologist forgets to total the heat units formed by additional cuts, the effect of the total accumulation on the tube is absolute: the tube always remembers.

Simultaneous Multilevel Tomography

Multiscreen "book" cassettes are used for simultaneous multilevel tomographic studies. These "books" hold from 3 to 7 pairs of intensifying screens, depending on the type of unit purchased. Intensifying screens are paired either $\frac{1}{2}$ cm or 1 cm apart, and are specifically matched to a particular type of cassette. The first screen pair generally utilizes slow- or medium-speed screens, and the last pair of screens is somewhat faster in speed.

The major disadvantages of the book cassette include:

1. *The most posterior pair of screens is the fastest speed combination, and therefore produces the poorest quality of detail.* For example, pair number 7 in the 7 cm book is 7 cm from the undersurface of the x-ray table. This would place the radiographic film between the pair of intensifying screens in position number 7, almost 3 inches from the tabletop, a violation of a basic rule of image geometry. Combine this increase in object-film distance with a pair of rather large-grain intensifying screens, and radiographic detail must suffer.

2. *It is impossible to control scattered radiation within the book cassette.* Radiographs made with this type of cassette generally have a gray look. The fogging of the radiographic image is partially due to scattered radiation from the screen base material and the interspacing materials which serve as a compression device for film-screen contact. After remnant radiation exits from an object under study and passes through the grid to be recorded on the film, there is no scattered radiation clean-up mechanism interspaced between

each pair of screens. *The greater the number of pairs of intensifying screens per cassette, the more difficult it is to control internal scatter.* Because of this difficulty, the use of the 5 to 7 screen book combination is frequently restricted to the lung fields or very thin anatomic areas.

The multilevel cassette became popular in the late 1950's when the radiation dosage scare was at its height. The book cassette, particularly in the 5 to 7 film range, does significantly reduce radiation dosage to the patient, but this reduction in dosage is often at the expense of film quality. For example, an increase of $2\frac{1}{2}$ times the mAs value over an individual film technic will usually expose a 7 film book. If the existing mAs value for a single screen technic is retained and a 7 screen book is substituted, an increase of 15 to 20 kVp is usually sufficient to achieve the same degree of film blackening.

Obviously, every type of book cassette works somewhat differently, and these factors are approximate and not documented for everyday use. When a book cassette is used, one occasionally experiences a lack of radiographic darkening in the more posterior 2 or 3 films. This occurs with low kilovoltage values. Examinations of the skull or abdomen under 70 kVp can be affected in this manner, although lower kilovoltage values can be used with the chest because of the more radiolucent nature of the lung.

Tube Protection. Utilization of a book cassette is easier on an x-ray tube as long as the initial exposure for the multilevel study is kept within the individual exposure limitations of the tube. In using a 7 screen book with 250 mAs at 70 kVp, a total of 17,500 heat units is generated. In using 100 mAs at 90 kVp, a value of 9,000 heat units is generated. Compare these heat unit values with the total heat units generated by 7 exposures, each requiring 100 mAs at 70 kVp (7,000 heat units per exposure)—an aggregate of 49,000 heat units. The lower heat unit values (9,000 to 17,500) of the book cassette technics are significantly less injurious to the x-ray tube than the heat unit totals (49,000) of 7 individual exposures.

Quality Comparison. If the simultaneous multilevel technic is acceptable as far as film quality is concerned, then there is another bonus in multilevel tomography in addition to the lowering of radiation dosage to the patient and the reduction of trauma to the x-ray tube. It is obviously easier to expose a single simultaneous multilevel series than it is to expose 7 individual cuts. *This saving in time is based on the assumption that the 7 cuts are actually necessary.* Often a 7 level book study is used to produce 7 radiographs of questionable quality when 2 or 3 individual sections of superior quality could be made of the same area.

Most book cassettes must be secured in position under an open-side x-ray table. The thinner 3 film cassettes (with 1 cm spacing) or the 5 film books (with ½ cm spacing) generally fit into existing Bucky trays.

See Chapter 1 for additional information on book cassettes.

Plesiotomography

A special book cassette with very closely matched intensifying screens spaced approximately 1 mm apart is used for plesiotomography.[11] This cassette holds 4 pairs of closely matched screens and is used in a conventional Bucky tray. Although studies of the sella turcica and optic foramina have been accomplished with this unit, it has been used primarily to demonstrate structures of the middle and inner ear.[14] Despite the controversy that exists over the thinness-of-cut capability of a linear tomographic unit (40 or 50 degree amplitude), many of the early plesio studies utilized unidirectional linear tomographic units. McGann popularized the technical aspects of plesiosectional tomography in America, using a 40 degree linear unit.[13] Some authors feel that a pluridirectional tomographic unit is a *must* to achieve proper thinness of cut.

The exact 1 mm spacing of the pairs of intensifying screens in the plesocassette has several advantages. Equidistant fractional spacing, (1 mm per level) would be virtually impossible to guarantee over a 4 level tomo-

graphic run for two reasons: (1) it is difficult for a patient to remain perfectly still for 4 individual exposures; (2) it is almost impossible to vary the fulcrum of most sectional units 1 mm at a time, for most of these devices are manufactured with ½ cm to 1 cm (10 mm) increments. A 1 mm variation in fulcrum height would require 5 perfectly equal spacing variations over a ½ cm range.

An interesting adaptation of this technic is available to the radiographic department that would like to attempt an occasional plesiosectional study when a plesiocassette is not available. If it is desired to perform this study with individual cassettes, it is possible to do so by carefully restraining the head and by using some method to circumvent the difficulties of fulcrum adjustment. This is not as complicated as it may seem. First, one must predetermine the lowest level desired in a 4 mm study. For example, let us choose a level at 9 cm from the tabletop in the AP position to make the initial section. Three sheets of aluminum, each 1 mm thick and similar to the filters used in radiographic tubes, are now required for the plesiosectional study. The initial 9 cm cut is made with a routine cassette, and then the first sheet of aluminum, exactly the size of the cassette in use, is inserted in the Bucky drawer and a 2nd cassette is placed on top of the aluminum sheet, elevating it exactly 1 mm from the position of the previous cassette. Fulcrum level is maintained at 9 cm, but the 2nd cut results in a 9.1 cm level. A 3rd cut is made with 2 sheets of aluminum in place and a 3rd cassette, resulting in a 9.2 cm level, and a 4th cut is made with 3 sheets of aluminum in place beneath a 4th cassette, resulting in a 9.3 cm level. The 4 tomographic cuts, 9.0, 9.1, 9.2 and 9.3 cm, are all made with a fixed fulcrum level of 9.0 cm. *It is very important that one use the same make cassette for all 4 exposures. Differences in the thicknesses of cassettes or variations in methods of screen mountings could destroy the effect of the 1 mm spacing.*

If an individual cassette is used for 3 or 4 exposures (for example, a 10″ × 12″ cassette divided into 4 equal 5″ × 6″ well collimated spots), the aluminum spacing technic can still be used. The restraining of the part under study must be stressed, since variations in patient positioning are more critical with 1 mm interspacing than with the conventional 1 cm (10 mm) cuts in general use. The slightest change can go unnoticed even by a highly trained technologist, disrupting the 1 mm spacing.

Great care with radiation dosage must be taken in performing plesiotomography. Tomographic studies in the vicinity of the cornea can possibly be instrumental in forming a cataract. The possibility of a cataractogenic dose to the human cornea must be considered whenever tomographic studies in or near the orbits are attempted. In utilizing a pluridirectional tomographic unit with a hypocycloidal movement, it has been reported that a dose of 10 R was delivered to the cornea when 8 exposures of each side of the skull were taken, for a total of 16 exposures. A simple lead shield placed over the cornea would have reduced the dosage to 1.0 R.[3]

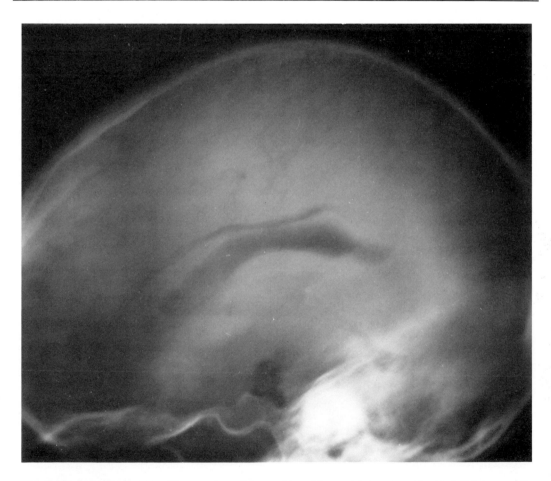

Fig. 8-13. Autotomogram. The moving of the head in a to-and-fro manner while in the lateral position during an air study of the ventricles of the brain results in an autotomogram. A gaseous contrast agent must be in the structure under study as the exposure is made. The 4th ventricle, aqueduct of Sylvius, and a portion of the 3rd ventricle are shown because they are in the midline of the skull or the axis of rotation. The dense petrous bones are obscured by the movement.

OTHER RADIOGRAPHIC PROCEDURES UTILIZING MOTION

Some radiographic procedures utilizing motion are difficult to catalog. They include:

Autotomography

Autotomography takes advantage of movements of a part of the body to blur out, in a body section fashion, superimposed anatomic structures. Some of the uses for autotomography are the making of the following studies:

1. **AP Cervical Spine.** By means of a talking-like motion the mandible is made to blur out so that the entire cervical spine is demonstrated from the 1st cervical vertebra to the last cervical vertebra.

2. **Thoracic Spine.** The patient is instructed to breathe gently throughout a long exposure. The rib markings and lung markings are blurred out by the use of the relatively long exposure time. This can be helpful in examining severely injured or emotionally distressed patients who cannot or will not respond to instructions.

3. **The AP Projection of the Lumbar Spine.** Gentle breathing throughout a prolonged exposure can be helpful to obscure superimposed gas shadows from the lumbar spine.

4. **Sternum.** A long exposure in the oblique position (5 to 10 seconds) is made during gentle breathing to blur out the ribs and lung markings.

5. **Pneumoencephalography.** A gentle rocking of the head to and fro in the lateral position after the injection of a small amount of air via the subarachnoid space during a pneumoencephalographic procedure will produce a tomographic-like effect, demonstrating the air-filled structures within the center of the brain. The 4th ventricle, aquaduct of Sylvius, and the 3rd ventricle can generally be seen, for they are in the center of rotation[5] (Fig. 8-13). This film is of great importance if a pneumoencephalographic study is being performed in a room that does not have a tomographic unit.

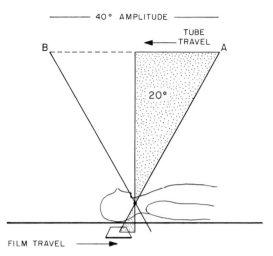

Fig. 8-14. Asymmetrical tomography. An asymmetrical exposure at the beginning or near the end of tube travel can help with certain studies to avoid the streaking artifacts associated with linear tomography. An *advanced asymmetrical* section of the larynx avoids the superimposed streaking of the central incisors and nasal bones. See Figure 8-15.

Asymmetrical Tomography

An asymmetrical exposure at the beginning of tube travel (*advanced asymmetry*) or near the end of an exposure arc (*retarded asymmetry*) can help to avoid some of the streaking artifacts associated with linear tomography in specific areas. McInnes, by using an advanced asymmetrical cut (the first 20 degrees of a 40 degree arc) to tomograph the larynx, avoids superimposed streaking of the central incisors and nasal bones[15] (Figs. 8-14 and 8-15).

Tomoscopy

A pluridirectional tomographic device with an image amplifier is available for tomoscopy.[4] Tomographic localization in the horizontal position is made with the use of the intensifier, and tomograms are quickly exposed (up to 4 exposures for each 10″ × 12″ film). Fluoroscopic guidance permits extremely tight primary ray collimation. Crysler claims that rarely does he utilize more than 3 minutes of elapsed time for the procedure.

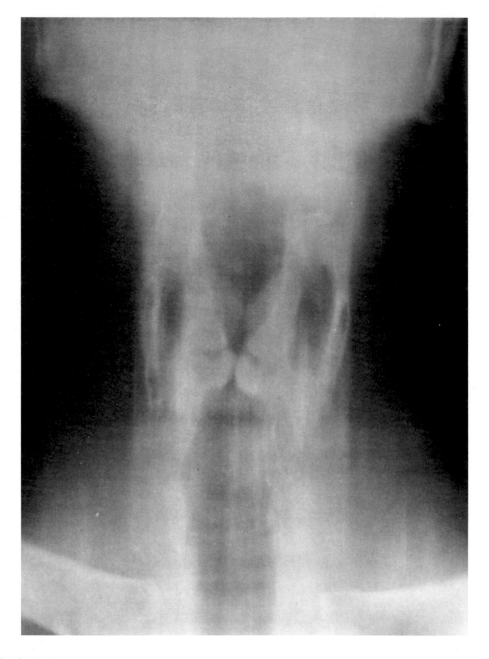

Fig. 8-15. Asymmetrical linear section of the larynx. A film is exposed with a linear tube movement (from the foot to the head of the patient) at a 40 degree exposure angle. The x-ray tube is energized for only the first 20 degrees of the tube movement. This relatively thick cut (20 degrees) made with an asymmetrical linear motion avoids the superimposed streaking of the central incisors and the nasal bones so often seen in a conventional symmetrical tomogram.

Horizontal or Axial Transverse Tomography

Horizontal or axial transverse tomography requires elaborate special equipment to create a section perpendicular to the axis of the patient. An axial transverse tomogram is obtained by simultaneous rotation of the patient and cassette about a parallel axis. These units are quite rare and are beyond the scope of this textbook (Fig. 8-16).

Pantomography

A panoramic body section view of the entire maxilla and mandible can be made with a special device that rotates or moves both the x-ray tube and the cassette during an exposure.[8,20] A narrow primary beam is used with a long exposure time (in some units, 20 seconds) to produce a tomographic image of the teeth (Fig. 8-17). There are many types of Pantomographic machines. A detailed description of their operation is beyond the intent of this textbook.

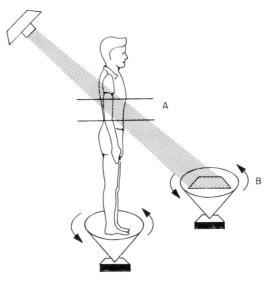

Fig. 8-16. Axial or horizontal tomography. An axial or horizontal tomogram can be obtained by the simultaneous rotation of both patient and cassette (B) about a parallel axis. A thick horizontal section (A) is the result of this technic.

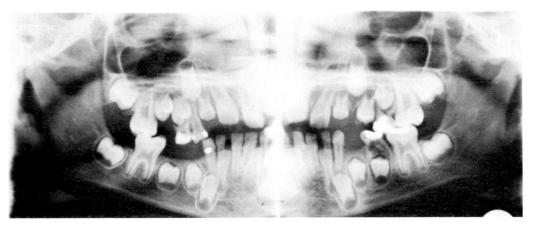

Fig. 8-17. Pantomography. A pantomographic unit uses a slit scan collimation technic. The tube and film are moved opposite to each other over a long slow but steady arc. A film is made of both the maxilla and mandible from one temperomandibular joint to the other.

REFERENCES

1. Andrews, J. R.: Plainography, introduction and history. Amer. J. Roentgen., 36:575–587, November 1936.
2. Bocage, A. E.: French Patent No. 536464 [1922].
3. Chin, F. K., Anderson, W. B., and Gilbertson, J. D.: Radiation dose to critical organs during petrous tomography. Radiology, 94(3):623–627, March 1970.
4. Crysler, W. E.: Tomoscopy and related matters. Amer. J. Roentgen., CIX(3):619–623, July 1970.

5. Cullinan, J. E.: Overlay: Anatomical Teaching Technique. p. 4. Wilmington, Del., E. I. du Pont de Nemours, n.d.
6. Cullinan, J. E., Jr.: Fractional focus x-ray tubes: Newer clinical and research applications. Radiol. Techn., 39(6):333–338, 1968.
7. Epstein, B. S.: Body-section radiography. Med. Radiogr. Photogr., 32(1):2–12, 1956.
8. Grigg, E. R. N.: The Trail of the Invisible Light. pp. 517–519. Springfield, Ill., Charles C Thomas, 1965.
9. Grossman, G.: Lung tomography. Brit. J. Radiol., 8:733–751, December 1935.
10. Korach, G., Vignaud, J., and Lictenberg, R.: Selective employment of Polytome in accordance with type of examination. Medicamundi (Eindhoven, Holland), 11(3):82–92, 1966.
11. Lapayowker, M. S., et al.: The use of plesiosectional tomography in the diagnosis of eighth nerve tumors. Amer. J. Roentgen., 88:1187–1193, 1962.
12. Littleton, J. T.: A phantom method to evaluate the clinical effectiveness of a tomographic device. Amer. J. Roentgen., CVIII(4):847–856, April 1970.
13. McGann, M. J.: Plesiosectional tomography of the temporal bone: A new multi-screen cassette. Amer. J. Roentgen., 88:1183–1186, 1962.
14. McInnes, J.: Tomography of the petrous temporal bone and optic foramina. Focus, 2(1):11–14, June 1959.
15. ———: Tomography today. Focus, 7(1):6–9, 1966.
16. Norman, A., and Wu (Ching), J.: Enlargement tomography in the diagnosis of bone tumors and skeletal trauma—a preliminary study. Bull. Hosp. Joint Dis., XXVII(1):46–50, April 1966.
17. Ring, J.: Elements of sectional radiography. Radiol. Techn., 38(1):17–22, 1966.
18. Seeman, H. E.: Physical and Photographic Principles of Medical Radiography. p. 114. New York, John Wiley & Sons, 1968.
19. Ibid., p. 118.
20. Smith, W. V. J.: A Review of tomography and zonography. Radiography, XXXVII(433):5–15, January 1971.
21. Stevenson, J. M.: Pleuridirectional body section radiography. The Focal Spot, 26(125):324–329, November 1969.
22. Svoboda, M.: Zonographic examination of the vertebrae. Rö-Blätter, 18:13–20, January 1965.
23. Updegrave, W. J.: Panoramix Dental Radiography. Dental Radiography, Rochester, N.Y., Eastman Kodak Company, 1963.
24. Weill, F.: The role of zonography in nephrotomographic technique. J. Radiol. Electr., 46:523–526, August-September 1965.
25. Weill, F., and Champy, M.: The use of zonography in cranio-facial radiology. J. Radiol. Electr., 46:77–82, January-February 1965.
26. ———: Further experiences with zonography of the maxilla and mandible. J. Radiol. Electr., 46:520–522, August-September 1965.

9. Rapid Serial Film and Cassette Changers

Modern angiographic technics require the use of high-speed film changers or cassette changers. (Fig. 9-1). It would be helpful if all fluoroscopic rooms could accommodate a rapid serial film or cassette changer. Since many fluoroscopic tables come equipped with movable tabletops that permit an overhang at either end of the table, conventional fluoroscopic rooms can be used as support rooms when repairs or renovations are underway in the angiographic room. It is not recommended that elaborate serial devices be utilized with routine fluoroscopic rooms, but the more simple 2 per second cassette or film changers can be wired inexpensively to any modern generator since many angiographic procedures can be performed at the rate of 2 exposures per second.

A distinct advantage of a film or cassette changer is that it uses a large-size x-ray film providing excellent detail resolution. Although cineradiographic studies serve a special function in the recording and demonstration of both anatomy and physiology, they do not approach the resolution potential of the screen radiograph. When a "serial" film study is being made, films are taken at predetermined intervals. When "rapid" is added to "serial" study, more than one radiograph per second is made.[4]

The making of additional films in a short time is shown to best advantage in the performance of coronary arteriography. Changers that operate at 6 or 12 frames per second are ideal for this type of examination. It is obvious that if more films are taken in a short period of time, more information becomes available (Fig. 9-2). It is also helpful

with specific examination if films can be made simultaneously in 2 planes, for biplane studies help to eliminate the problems created by chamber or vessel overlap. They also permit vessel or structure identification in both planes in a sort of 3-dimensional effect. Since both planes of the angiographic study are made simultaneously, the utilization of a single dose of contrast material is an added advantage. Smaller total amounts of opaque media are possible with a biplane procedure. Unfortunately, when biplane studies are performed, technical difficulties more than double. The selection of proper grids and collimating devices becomes critical.

Certain problems and difficulties are common to all serial film or cassette changers. With the exception of the roll film changer, films are frequently not in order when delivered to and from the automatic processor, and sequence marking can be a problem. Roll film must be cut to field size if it is to be stored in a conventional radiographic envelope. There is excessive waste with almost all film sizes, particularly if tight collimation is employed with a selective or subselective catheterization technic. For example, often when a single kidney is evaluated by selective catheterization, only a small segment of the film is used.

The earliest practical roll film device was that manufactured by the Fairchild Instrument Corporation. The basic mechanism of this roll film device approximated that of the aerial camera used by reconnaisance planes during World War II. The camera can advance large sheets of photographic film at the rate of 2 frames per second, with a remarka-

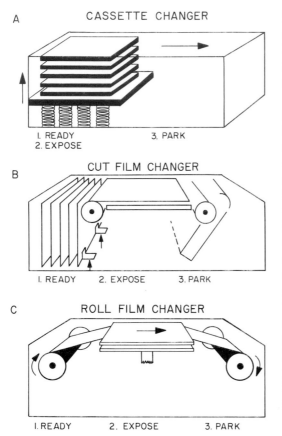

Fig. 9-1. Basic principles of serial changers.
Although serial film and cassette changers serve similar functions, their basic mechanical concepts are quite different.

(A) *Cassette changer.* The most popular cassette changer holds 12 (11" x 14") cassettes and may be operated as fast as 2 frames per second down to 1 frame per 2 seconds. The cassettes have a thick layer of lead beneath the back intensifying screen to prevent radiation leakage to the underlying cassettes. The cassettes are placed one on top of the other and are elevated by a spring-operated tray into the expose position. A simple chain mechanism pulls the exposed cassette into a park chamber, and the cycle is repeated.

(B) *Cut film changer.* The most popular cut film changer is loaded with a film holder (up to 30 sheets of 14" x 14" film) and can operate at a maximum speed of 6 frames per second. Two tiny ports at the bottom of the loaded film magazine are used to admit the ejection fingers of the unit, which flip the film upward into the receiving portion of the opened cassette. The film is conveyed into the cassette, the screens

bly good screen-film contact because of its smaller frame size, approximately 9" × 9".[10] It is only with the film changers of larger frame size that screen-contact difficulties are encountered.

Over the years progress has been made from the smaller frame, slow roll film changer to the 6 per second, large frame, variable-speed film changer system. This device includes a film changer, program selector, and a mounting stand. The major advantage of this apparatus is that it has removable film magazines for daylight loading and unloading. A biplane roll film changer is available with an 8 per second or 12 per second maximum capability, but a distinct disadvantage of this system is that both units are permanently linked together in a L-shaped form. The individual units cannot be separated or moved in any direction for tube-angle technics. The major disadvantage of this fixed biplane unit is that film loading must be performed in the x-ray room with the room lights off.[3] This can be a problem when additional films are needed.

TYPES OF SERIAL FILM OR CASSETTE CHANGERS

There are several types of film or cassette changers. These include the following:

Cassette Changers

The 2 cassettes per second (12 cassette maximum) changer has been available for

are compressed, and an exposure is made. The cassette opens, and the film is transported into the park position (the receiving magazine).

(C) *Roll film changer.* Two major types of roll film changers are in use today. The more popular unit is a 6 frame (11" × 14") per second maximum device with a 50 ft roll of film capability. The second unit, a biplane unit, can be purchased with an 8 or 12 frame (14" × 14") per second maximum, using an 80 ft roll of film. The film transport mechanism advances the film to the open cassette, screen compression is applied, an exposure is made, the cassette opens, and the film is advanced to the take-up roll.

many years. These cassettes are rather heavy, and the unit is quite noisy as the cassettes are shuffled from the load-firing area to the park area. Each cassette has a thick layer of lead beneath the back intensifying screen to minimize exposure leakage to the underlying cassettes. The cassettes are placed one on top of the other and are elevated by a spring-operated tray into the firing position (Fig. 9-1A). The shuffling of these heavy cassettes creates a vibration in the unit, and this vibration is the subject of widespread technical criticism.

The unit actually has an excellent dampening effect and can be used quite adequately without motion interference if a few simple rules are followed. A major complaint is that films of the skull in the AP position seem relatively unsharp. Most technologists will admit that the lateral radiographs made with this unit are acceptable, but that the AP radiographs exhibit motion. The unit works equally well in both positions, but a common technical error is made in the AP mode. Since most of these units come equipped with a plastic strap, the skull is generally strapped directly to the changer. By strapping the patient to the changer to depress the chin, we can create a motion artifact. When the first film is exposed, the cassette leaves its expose postion and falls in the park chamber. This rapid shifting of the heavy cassette causes the entire unit to vibrate. The unit and the strap vibrate, and the patient's head in turn vibrates. The unit quickly dampens, and motion virtually ceases. Unfortunately, the patient's head can still be moving. *The 2 per second cassette changer of this type should never be used as support for the head during cerebral angiography.* A radiolucent extension board should be used with the radiographic table to support the head independent of the cassette changer. The changer is then raised into position almost in direct contact with the head. Any attempt to immobilize the head should not include the plastic strap attached to the cassette changer. Restraining devices should be attached to the extension device of the radiographic table.

If large-field radiographs are made with

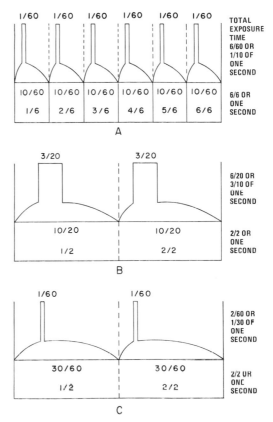

Fig. 9-2. Exposure rate per second. Some angiographic studies must be made at a rate of 6 to 12 frames per second if adequate diagnostic information is to be obtained. It is obvious that if more films are made in a shorter period of time, more information becomes available for evaluation.

When 6 films per second are made with an exposure length of $\frac{1}{60}$ of a second, the total exposure to the patient is $\frac{6}{60}$ of a second or $\frac{1}{10}$ of a second (A). The delay between individual frames is $\frac{9}{60}$ of a second.

When 2 frames per second are made with an exposure of $\frac{3}{20}$ of a second, the patient is exposed to a total of $\frac{3}{10}$ of a second per 1 second (B). The delay between exposures is $\frac{7}{20}$ of a second or $\frac{21}{60}$ of a second. As compared with A, this delay between frames is $2\frac{1}{3}$ times longer. If an exposure value longer than $\frac{3}{20}$ of a second is used, vessel, organ, or patient motion can occur.

If an exposure value of $\frac{1}{60}$ of a second is used to overcome motion in a 2 per second frame study, the increased time delay between frames is $\frac{29}{60}$ of a second (C). Valuable diagnostic information can be missed.

this device, only the exact number of cassettes required for the study should be loaded into the changer, for when each exposure is made, there can be a slight penetration of the next cassette by the unattenuated primary beam. When the AP skull projection is made on film 1, part of film 2 can be exposed by the unattenuated x-ray beam. If film 2 were to be developed, a complete outline of the skull would be seen; but the area where the Towne projection of the skull would be recorded is crystal-clear, for the skull acts as a primary beam attenuator, preserving the sensitized emulsion of the following frame. This faint outline of the skull is of no technical concern as long as the same projection is maintained. Frame 2 is exposed, frame 3 receives an outline of the skull, and so on through the entire serial run. If 12 cassettes were to be loaded in the exposure magazine,

and 6 used for the AP projection, then when the unit is turned into the lateral projection, the 7th film will have an outline of the Towne projection of the skull. The lateral borders of the blackened AP film will superimpose on the lateral projection of the skull, producing an artifact. It is important to use only the proper number of cassettes required for a serial run. When 2 different projections are made with a fully loaded (12 cassette) changer, the 1st cassette of the 2nd series should be discarded to avoid this superimposition artifact, or it should be removed from the changer and reloaded with unexposed radiographic film (Fig. 9-3).

Cassette Changer Using Vacuum Screen Contact Principle

An x-ray cassette changer utilizing 21 14" × 14" cassettes is available. These cassettes can be loaded with smaller (8" × 10" or 10" × 12") films for pediatric angiocardiography or selective arteriography. Instead of using the typical rigid metal cassette which relies on some type of hinge or lock to achieve mechanical pressure, the vacuum cassette enlists atmospheric pressure to produce almost perfect screen-film contact. The intensifying screen, film, and lead backing for the cassette are placed in the open end of a black plastic envelope which has been completely sealed on 3 sides. A special vacuum sealing unit is used to create a perfect vacuum, completely sealing the bag in about $4\frac{1}{2}$ seconds. This vacuum can be maintained over a very long period. The loaded cassettes are placed in the changer, one on top of the other, and as each exposure is made, the cassettes are elevated slightly. A push bar moves forward, thrusting the exposed top cassette into a pair of rollers, which draw it into the adjacent receiving chamber. This process can be repeated up to 21 times, and may be used as often as 4 times per second. Since the unit is both large and heavy, it has an overall rigidity which helps to eliminate vibration. The top of the device is easily opened, and grids of different sizes or ratios can be interchanged. This feature can be of

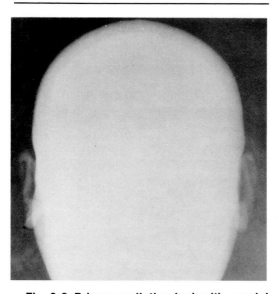

Fig. 9-3. Primary radiation leak with a serial cassette changer. When a serial cassette changer is used for cerebral angiography, there can be a leak of primary or unattenuated x-ray to the underlying cassettes. This faint outline of the skull is of no technical concern unless the position of the head is changed—for example, to the lateral projection—for a second serial exposure. It is therefore important to load the unit with the exact number of cassettes required for each serial exposure.

significant value when it is compared with serial changers having built-in grids.

Any make or speed or combination of intensifying screens can be utilized with the vacuum cassette changer. It can be loaded with fast-speed screens, for example, to assure an adequately exposed abdominal series. The tabletop could then be moved for a femoral artery study, and the cassettes used could house medium-speed screens. The tabletop could then be moved again to record the arteries of the distal extremities down to and including the ankle, and a single medium-speed screen could be used for a 50 per cent reduction in film density, adequately exposing the films of the lower extremities. The cassettes can be loaded with films of different speeds. The thicker abdominal area requires an extremely fast film, whereas a slower film for the thinner lower extremities could be used.

This unit, in connection with a computer programing device, can be loaded with the same screen-film combination, and every phase of the operation can be controlled automatically. Exposures can be made, power syringe triggered into operation, and the tabletop moved automatically by means of an elaborate programing device. A punch card is used to designate any action or interval required for the angiographic study. The punch card is inserted into the programer, and the examination is initiated by the technologist. Exposure times can vary, frame intervals can vary between exposures, and kilovoltage can be raised or lowered to preselected values as thicker or thinner segments of anatomy are presented for study.

Cut Film Changers

Cut film changers are available in multiple film sizes, the most popular being 14″ × 14″ or 10″ × 12″. The magazine of the most popular cut film changer can be loaded with a maximum of 30 films; an elaborate programing device can be used with this unit, which can operate as fast as 6 films per second (Fig. 9-1B). Operation of the numbering and marking system of this unit is inde-

pendent of the collimation of the x-ray beam.

Basic Mechanical Operation of the Cut Film Changer. This unit has a magazine in each plane which holds 30 sheets of x-ray film, each film being separated by a stainless steel wire frame. The magazine is preloaded in the darkroom, its lid and ports are closed, and it is inserted into the load portion of the film changer. The receiving cassette of the unit is set in position. The tray on which the magazine rests is slowly advanced electrically. The lid of the magazine, which is retained by a locking mechanism, is automatically opened as the magazine moves forward. Two small ports are simultaneously opened at the bottom of the magazine, and during the operation of the unit, 2 levers enter through these holes and flip a single film upward. The film is conveyed through the system into the opened cassette, the cassette closes, an exposure is made, the film is emptied into a receiving magazine, the ejection fingers flip another film into place, the screens close, an exposure is made, and the cycle is repeated.

Screen Contact of the Cut Film Changer. A unique method is used to improve screen-film contact with the 6 per second cut film changer. The lower intensifying screen is mounted on a slightly curved plate. The center of the lower intensifying screen, therefore, strikes the center of the upper intensifying screen rather than the entire screen. Theoretically, this premature closing of the screen in the central portion expels air from the sides of the cassette, helping to ensure better screen contact. Screen contact in the center of the films produced by this film changer shows definite improvement, particularly in the larger 14″ × 14″ version. There is, however, some questionable contact on the peripheral segments of the films.

Roll Film Changers

The most popular roll film changer is available as a single or biplane unit and is capable of 6 exposures per second in each plane. A program selector is available with this unit which can accept a roll of radiographic film 14 inches wide by 50 feet long (Fig. 9-1C). Approximately 50 11″ × 14″ radiographs are possible with this unit, although there is the inevitable loss of film when scout films or short serial studies are made, and a new leader has to be applied to the receiving magazine.

An elaborate roll film device manufactured only as a biplane unit and capable of 12 films per second in each plane is available. This unit holds approximately 75 frames in the vertical plane, and 65 frames in the horizontal plane. It is possible to use rolls of film up to 80 feet in length. An excellent programing device permits an exposure frequency from 12 films per second to 1 film in 10 seconds. The unit is also available with a maximum exposure frequency of 8 radiographs per second.

Combination Roll-Film-Cut-Film Changer

A new unit has been recently introduced with an 8 frame per second capability. These units are separate changers and can be used for either single plane or biplane operation. A roll of film can result in 140 14″ × 14″ cut films. Maximum capacity of the exposed film magazine is 40 films. Roll film is used for the loading of the changer, but as the roll film enters the receiving magazine, it is cut into individual 14″ × 14″ frames. This system has several benefits, including the simplicity of roll film loading and transportation versus the tedious loading of a cut film or cassette changer. Also, an unusually large amount of film is available: 140 14″ × 14″ frames. The availability of cut film in the receiving magazine is helpful, for only a few frames of the study need be processed for a diagnostic or technical check. When a roll film changer is used, the entire roll must be processed before a specific film can be seen. If the roll-film-cut-film changer were used for a study such as cerebral angiography and 12 films were made per injection, the 1st or 2nd film of a study could be processed to determine adequate arterial filling, and the final film could be processed from the standpoint of adequate venous filling. After the darkroom technician removes the 1st, 2nd or last film from the magazine, he could continue to process the remainder of the study, but not at the expense in time of the angiographer or the patient. A complex programing device is available with the roll-film-cut-film changer.

See-Through Changers

"See-through" film changers are used in combination with an image intensifier.[1] It is no longer necessary for a patient to be moved after fluoroscopic placement of a catheter from the angiographic table to a position over the film changer. The see-through film changer has a large area in the center that permits the use of an intensifier in tandem with the changer. This opening facilitates patient positioning as well as collimation of the primary beam and permits the technologist to observe the angiogram "live" on a television monitor. If an angiographic examination is performed with a video tape or video disc recorder, images can be immediately replayed and evaluated.

A cut film changer holding up to 20 films (14″ × 14″ in size) is available with this see-through feature. This unit is extremely compact, and can be used as a frontal or lateral unit or mounted in the biplane mode in areas that were considered inaccessible for a film changer. Several x-ray manufacturers or private investigators have see-through changers in use as either prototypes or 2nd- or 3rd-generation experimental units. The commercially available see-through changer utilizes a punch card control system for actuation of the power syringe, programing, number of films per second, and so on. The use of the see-through changer should increase, for as angiographic procedures become more commonplace, the time loss in processing large volumes of x-ray film for immediate viewing can no longer be tolerated.

BIPLANE CHANGER OPERATION

Many feel that the use of a film or cassette changer in a single plane is inadequate for many studies, although many angiographic laboratories utilize single-plane changers. Multiple injections of opaque in several positions are performed to achieve a biplane effect.

Unfortunately, when a decision is made to graduate from single plane to biplane, technical problems more than double, being compounded in a way that defies a mathematical evaluation.

Scattered radiation becomes the most significant technical difficulty in working with biplane angiographic technics. Although the use of cross-hatch grids with perpendicular beam technics is recommended for biplane studies, it is impossible to use a cross-hatch grid in the frontal plane during a cerebral angiographic procedure. A linear grid is used for the AP projection of the skull, whereas a cross-hatch grid is frequently used for the lateral projection. The scattered radiation generated in the AP and lateral planes is easily absorbed by the cross-hatch grid in the lateral plane. Unfortunately, the linear grid in the AP plane is often not quite able to absorb the scattered radiation generated by both planes. Particularly disturbing is a density increase across the AP plane created by the lateral beam. For example, let us assume that the patient is supine in the left lateral position; that is, the patient's head is over the AP changer, and the left side of the head is against the lateral changer. The x-ray beam enters the right side of the patient's head and exits from the left side to expose the film in the lateral film changer. The AP radiograph will exhibit a decreasing wedge of density from the right to the left side of the film. The placement of a primary beam attenuator on the outer aspect of the frontal changer can help to eliminate this density artifact (Fig. 9-4).

An adjustment of the technical factors may be necessary to overcome the enlargement of the skull in the lateral projection due to the position of the lateral changer. Frequently the unit cannot be placed against the patient's head in the lateral projection due to the width of the shoulders or the shortness of the neck

of the patient. This creates an increase in the object-film distance with object magnification. To overcome this difficulty, the lateral tube is moved to a 48 inch or a 72 inch focus-film distance, and an increase in technical factors must be made. A generous increase in kilovoltage can be tolerated by the cross-hatch grids in the lateral projection. This increase in kilovoltage, with a corresponding increase in scattered radiation, unfortunately cannot be tolerated by the linear grid in the AP plane. Every effort should be made to ensure good patient contact with the lateral film changer to avoid the excessive kilovoltage values necessitated by an increased focal object distance.

Scattered radiation can be avoided in operating a 6 per second cut film changer by the *alternate loading of the film magazines.* Since each magazine holds 30 frames per unit, the magazine can be loaded in the following fashion: The frontal unit can be loaded with a film in frames 1, 3, 5, 7, etc., and the lateral unit can be programed with film in frames 2, 4, 6, 8, etc. The unit is activated, film 1 is exposed in the AP frame, film 2 is then exposed in the lateral frame, film 3 in the AP film, film 4 in the lateral, and so on. Thus the biplane scattered radiation difficulty is eliminated by alternate loading.

Some roll film changers can be operated with alternate exposures during a biplane study. Although scattered radiation is avoided, there is a significant economic disadvantage, for, since the units utilize roll film, both rolls will simultaneously advance, and as alternate frames are exposed, the corresponding frames in the opposing projection are wasted.

Horner describes an unusual biplane configuration using 2 roll film changers for cervical arteriography.[6] The patient is placed in the supine position, with the head and neck resting on an extension of the tabletop. Both changers are angled to form a "V," and both radiographic tubes are positioned perpendicularly to their respective changers. Simultaneous oblique projections of the arteries of the neck and skull can be made with a single injection of contrast material. A strip of alu-

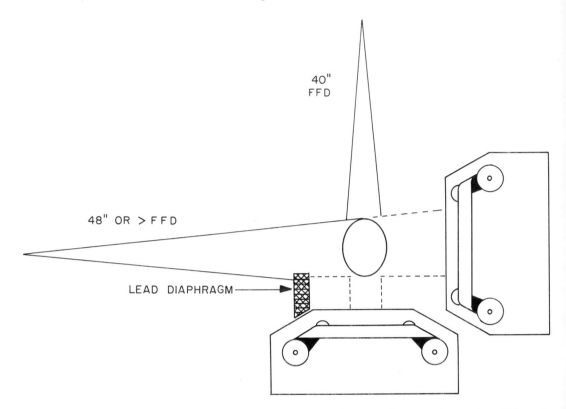

Fig. 9-4. Biplane scattered radiation difficulties during cerebral angiography. Although cross-hatch grids can be used with the lateral projection during cerebral angiography, a linear grid is required for the AP (tube-angled) projection. Often there will be a disturbing increase in film density across the AP view, decreasing from the tube side to the lateral changer.

Placing a *primary* beam attenuator on the tube side of the frontal changer helps to eliminate this artifact. The frontal series is spared the effect of the primary beam. The lead must be carefully placed to avoid the cutting off of a segment of the skull in the lateral position.

minum (2 mm thick) is placed on each changer and acts as an underpart filter to attenuate the x-ray beam in the thinner cervical area. The unfiltered remnant beam records the more dense area of the skull for a well balanced radiograph.

Kilovoltage Adjustments for Biplane Technics

Regardless of the type of collimator or grid used, there still is a considerable increase in film density when a biplane procedure is performed. In general, if an adequate biplane study has been achieved, and a single plane repeat examination must be remade, an increase in technical factors is needed to overcome the density effect of the scattered radiation generated by biplane operation. On occasion a biplane examination will be made of the abdominal aorta, and the lateral projection will be utilized only during the aortic fill stage. The films in the biplane mode are technically adequate until the study reverts to single plane operation. When the biplane study is adequate, the AP single plane is generally light. When the AP single plane is adequate, the AP films made in a biplane mode are frequently dark.

Biplane Operation With a
Single Generator

The most common error made in purchasing angiographic equipment is the sharing of a single generator to accommodate 2 serial film or cassette changers. Although the splitting of a single generator can be accepted for some angiographic procedures, in general it should be avoided. As 1,000 mA and 1,500 mA units become available, the splitting of a generator is theoretically acceptable but must be determined by the nature of the procedures contemplated.

When a biplane split is made with a single generator, both tubes are frequently required to operate at the same kilovoltage level and at the same exposure times. This problem can be difficult to overcome. For example, if a biplane split were made to examine an abdomen, it is difficult to balance the AP and the lateral exposures. The lateral projection of the abdomen requires considerably more radiation for adequate film blackening than does the AP plane. Fast-speed screens are often needed to achieve adequate film blackening in either plane. For example, if the technologist were to use a 600 mA generator and decided to split this unit for biplane operation (300 mA per plane), with a film changer having a maximum exposure angle of $\frac{1}{60}$ of a second at 6 frames per second, only 300 mA at $\frac{1}{60}$ of a second (5 mAs) would be available for each plane. It is questionable that a 6 frame per second run, at a full output of 600 mA at $\frac{1}{60}$ of a second (10 mAs) would be acceptable for most angiographic studies. Too often a generating apparatus is purchased which is inadequate for the *maximum exposure length per cycle* capability of the serial changer to be used (Fig. 9-5). This problem is compounded by the splitting of the generator for biplane operation.

Simultaneous biplane operation with mA station distribution is available. The total mA values selected for the 2 tubes cannot exceed the output of the x-ray generator. For example, if a unit is available with 1,000 mA total capability, it can be split with 600 mA for

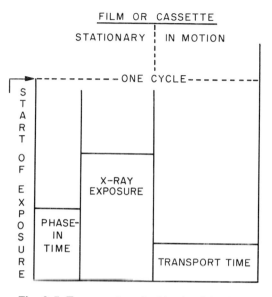

Fig. 9-5. Transport cycle. Much of the frame per second cycle is of little value to actual exposure length. The closing of the relays, necessary to fire the x-ray machine, cassette compression, and film or cassette transport, all rob the angiographic technologist of usable exposure time. Relays can be adjusted to eliminate delay. There is, however, a delay prior to the exposure known as "phase-in time." This delay can vary with different types of generators from 0 to 16 milliseconds. Phase-in time can be particularly disturbing when a 6 or 12 per second series is planned.

---

the lateral plane and 400 mA for the AP plane, or vice versa. From the standpoint of radiation output, it is still important to stress that the design of most biplane rooms leaves much to be desired, even when 2 generators are utilized with 2 serial changers. *It is important to determine output needs long before equipment installation begins. The output of new equipment must be weighed against the maximum exposure length per cycle of the serial unit contemplated.*

SOME ASPECTS OF
SERIAL PROCEDURE TECHNIC

Heat Unit Considerations During
Serial Angiographic Procedures

When serial angiographic studies are performed, the capacity of the x-ray tube can be a serious limitation. Focal spot size, as well as the heat storage capacity of the anode, must be seriously considered.[2] The cooling rate of the tube housing is of some concern. When multiple studies are attempted, cooling time for the x-ray anode should be included. If a second angio run is attempted before proper cooling of the anode has taken place, even though the total heat unit value of the run is acceptable, tube damage can occur.

Some Technical Difficulties Common to Serial Changers

The purpose of the serial film changer is to furnish as many frames per second as possible during an angiographic study while an attempt is made to stop completely the motion of pulsating arteries or vessels. This is probably not totally possible, for the sharp recording of pulsating vessels is determined by several things: the motion of the vessel and the patient, the motion of the serial film changer itself (some units exhibit tremendous vibration while in operation), the accuracy of the screen-film contact mechanism of the system, or the dependability of the film or cassette transport system.

By far the most significant problem encountered by the angiographic technologist is the poor screen contact encountered in many changers. With the exception of the vacuum cassette changer, no unit achieves the screen-film contact of conventional cassette radiography. This is not to say that screen-film contact is not adequate in serial changers, for some of the cassette changers achieve excellent screen-film contact, but one must remember that screen-film contact must be maintained in a moving mechanical system. Inherent vibration can create a motion effect, with corresponding image unsharpness.

Jamming Possibilities

Jamming possibilities with cassette changers are due to the moving of large masses of weight. The roll film changers can also jam, generally because of poor mounting of the film on the receiving or loading spools. The

6 per second cut film changer will jam readily if the manufacturer's recommendations are ignored. Adjusting the levers that flip the film into position in the cassette is critical, and a special measuring tool is available to assure their proper position. To avoid cut film becoming caught in magazine jams, it is advisable to leave a single sheet of unexposed film in the receiving magazine. This allows the first exposed film to pass easily into the receiving magazine, avoiding any undue resistance from film-metal contact.

It is very important that with serial cut film changers an antistatic agent be used to prevent static artifacts. Silicon coatings are available that can be applied to the intensifying screens, but these sprays must not strike the rubber rollers and so cause them to lose their traction effect which is necessary for the transport of films. Available for serial film changers are special films that are made with some form of silicon or teflon additives. These additives help to ensure proper film transport with minimal difficulties.

Automatic Film Marking

It is important that some type of device be used to number films consecutively during a serial angiographic study. There is a device that optically transfers pertinent information, such as the name of the patient, type of examination, date, etc., directly onto the radiographic film. This system is completely independent of any form of x-ray collimation, which is a *must*, for as selective angiography is performed, smaller segments of film are exposed to radiation. Regardless of the size of the x-ray beam, the optical light system will accurately mark each film in a consecutive fashion.

Type of Intensifying Screens To Be Utilized in Serial Film Changers

Medium-speed screens, as a general rule, should be used with fast-speed radiographic film to avoid the problems encountered with the thicker but fast intensifying screens. Since fast-speed screens are manufactured with an increase in the layer thickness of phosphors, there is an increase in the halation effect. Very small objects of low contrast, such as extremely small vessels, can become very difficult to evaluate. In employing tube-angle technics, medium-speed screens produce sharper images because the parallax effect is minimized.[8] For conventional angiographic procedures in which right-angle beams are utilized in both the AP and the lateral projection, fast-speed screens and fast films can be used to overcome motion. The use of fast-screen-fast-film combinations with moderate or low kilovoltage values produces a very definite improvement in contrast. In the near future it will be possible to utilize medium-speed screens routinely with high mA output units and relatively short times (hopefully in the millisecond range).

Regardless of the theoretical possibilities of screen and film combinations, the output of the equipment is the limiting factor. If an angiographic room is limited to a 500 mA single-phase generator with a serial film changer having a maximum exposure length per cycle of $\frac{1}{60}$ of a second (at 6 frames per second), it is quite obvious that fast screens and/or fast films are required. *Film blackening as well as the elimination of motion, must be accomplished within the limitations of the equipment and accessories* (Fig. 9-5).

It is important to keep the intensifying screens of any film changer clean. Cleaning should be performed at least 2 to 3 times per week, depending on the activity in the room.

Changing Intensifying Screens in a Serial Film Changer

The complexity of new types of generating equipment is often stressed, whereas the basics of radiologic technology are frequently ignored, and entirely too much emphasis is put on the changing of equipment design. For example, a new type of serial film changer would be useless without intensifying screens, or without a proper collimating device or grid. And yet it is remarkable how great a problem develops when one attempts to change intensifying screens or a grid in

an existing serial unit! A set of intensifying screens for a roll film changer can be professionally installed for approximately $100, a remarkably low price in consideration of the cost of many of these changers ($10,000 or more). Yet many serial film changers in use today are producing films of poor quality because of the use of old intensifying screens, a poor collimating device, an initial improper grid selection, or an improper screen-film combination.

Intensifying screens are used for extended periods of time without regard to phosphor damage, intensification fall-off, etc. Although obvious screen artifacts as well as poor screen contact are quickly noticed, screen gain can deteriorate in a gradual and unnoticed manner. Unfortunately, a gradual compensation is made for this light loss by an increase in x-ray factors. Increased exposure to the patient as well as the lengthening of exposure time should be a matter of critical concern, for motion blur can completely destroy smaller vessel images.[7]

A busy angiographic room can utilize 30 or more rolls of radiographic film per week, at approximately $30 per roll. If one were to estimate the total cost of roll film plus processing, a figure of $1,000 per week would be fairly conservative. Over a year the investment in film and chemistry in a typical angiographic room could easily total $50,000. This figure does not include such costs as opaque media, equipment deterioration, special surgical trays, the salaries of technologists and angiographers, etc. But $50,000 represents only the cost of the film and its processing. Although $50,000 a year is spent for film and chemistry, how difficult it is to justify the installation of a pair of intensifying screens several times per year! Even when the cost of the screens is not a problem, the unconvincing argument is given that a room cannot be spared for the time that is required to change the intensifying screens.

Grid Selection

The selection of a proper grid for the serial changer should not be left entirely to the manufacturer's representative. Equipment output and type of examinations, as well as the type, focus, and ratio of the grid, etc., should all be weighed before a final selection is made. In the opinion of this author the interspacing material used in the grid is extremely important, and the use of the aluminum interspaced grid is recommended for the reasons documented in Chapter 2.

When the technologist is operating in a single plane, it is also recommended that a cross-hatch grid combination be utilized rather than a high-ratio linear grid. For example, many changers are equipped with a 12 to 1 linear grid with a focus-film distance of "exactly" 40 inches. This means that in any study attempted with this unit the technologist would have to utilize a 12 to 1, 40 inch focused grid whether the study happened to be a pediatric angiogram or an abdominal aortogram in the lateral position—too much grid for the infant, and not enough grid perhaps for the larger patient! If an 8 to 1 aluminum interspaced grid were used as the basic grid for the serial changer, then a 2nd overlay grid could be purchased—a 5 to 1 grid with aluminum or fiber interspacing. In the experience of this author, fiber interspaced grids in the cross-hatch position are not objectionable. Apparent defects in one grid are lost in the weave pattern created by the other grid. It is acceptable to use a linear aluminum interspaced grid with an overlay linear fiber interspaced grid. If the aluminum interspaced grid is 8 to 1 in ratio, it is possible to use this grid at a focus of 34 inches to 44 inches. If the overlay grid is a 5 to 1 grid focused at 28 to 72 inches (beyond the focus capability of the 8 to 1 grid) when the the grids are used in combination, the focus of the higher ratio grid (8 to 1) is of primary concern for focus-film length determination. This means that a combination has been achieved of 8 to 1 and 5 to 1 in a cross-hatch mode, equal to or better than the capability clean-up of a 12 to 1 linear grid, trading a fixed 40 inch focus-film distance for a variable 34 inch to 44 inch focus-film distance. In return, the

ability to angle the x-ray tube for such technics as cerebral angiography in the AP projection has been given up.

What are some of the possibilities with the cross-hatch combination previously recommended? First, the aluminum interspaced 8 to 1 grid can be used without the overlay grid for tube-angle technics. With proper collimation of the primary beam, an 8 to 1 grid is more than adequate for cerebral angiography. The 34 inch to 44 inch focus-film range helps the technologist to overcome focus-film length variations encountered when excessive tube-angle technic is needed to overcome positioning difficulties (up to 45 degrees)—for example, positioning people with short necks or a severe kyphosis. Second, the linear 8 to 1 grid is more than adequate for pediatric angiography, lowering dosage to the infant as well as significantly shortening exposure time, as compared with the fixed 12 to 1 linear grid. Third, again patient dosage and exposure times are lowered during adult peripheral arteriography or venography of the upper or lower extremities, as compared with the 12 to 1 linear grid. Fourth, the cross-hatch combination of the 8 to 1 and 5 to 1 grid can be utilized for angiograms of large patients, and is quite effective in the biplane mode when tube angles are not required. Fifth, it is possible in attempting selective angiography, such as direct placement of a catheter into the renal artery, to use the linear grid without the overlay grid with severe collimation of the primary beam. The primary beam can be restricted to the body of the kidney so that there is a significant decrease in the generation of scattered radiation, making the work of the linear grid easier. It is also possible, by careful collimation of the primary beam and the removal of the overlay grid, to increase the number of frames per second during a study. A slight increase in the kilovoltage value coupled with the removal of the overlay grid could result in a 4 frame per second examination in larger patients as compared with a 2 frame per second examination.

The use of selective collimation in conjunction with a cut film changer has been shown in an article by Franji.[5] A serial study of a small infant or a small adult organ, such as the kidney, can be exposed on one half of the 14″ × 14″ film. The exposed films are then removed from the receiving magazine, reloaded in the loading magazine in the darkroom, and reused for the 2nd projection. The part under study is centered to the opposite or unexposed side of the 14″ × 14″ film, and another examination is made.

When a serial changer has an externally mounted grid, a major technical problem can be overcome. For example, in using biplane 14″ × 14″ serial changers, it is extremely difficult to maintain good object-film contact in the lateral position, for if cerebral angiography is being attempted, the skull must be centered to the 14″ × 14″ grid. This position automatically produces a 3 or 4 inch gap from the lateral aspect of the skull to the lateral changer. When the lateral serial film changer is moved into position, there is a significant increase in the object-film distance of the skull in the lateral projection. If the technologist were to use a 10″ × 12″ grid in the AP changer placed on the side of the changer nearest the lateral film changer, a decrease in the object-film distance would result (Fig. 9-6).

Film Processing

In a busy angiographic laboratory, an automatic processor should be available. Approximately 50 films must be processed daily to prevent self-exhaustion of the solutions in an automatic processor.[9]

If a rapid-process 90 second unit is used with appropriate 90 second chemistry in an angio area or any other independent duty area, it is important that an adequate amount of film be processed to replenish the unit properly. When these units are used on a limited basis, they have a tendency to self-exhaustion of the solutions. If the angiographic volume is decreased, the unit should be utilized for conventional film processing. Routine studies should be developed in the standby processor to maintain proper chem-

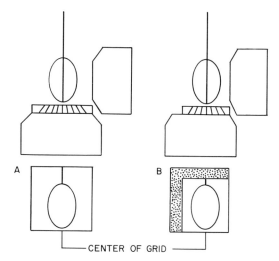

CENTER OF GRID

Fig. 9-6. Cerebral angiography with a large frame serial film or cassette changer. When the skull or other small body part must be centered in the AP position to a large frame (14″ x 14″) film or cassette changer, an increase in OFD occurs in the lateral position. Using a smaller grid (10″ x 12″) moves the center of the grid off center toward the lateral changer. The patient can be moved closer to the lateral changer, and thus a significant reduction in OFD is achieved.

ical balance. If the workload volume is low in the angio area, special attention should be given to the mixing of the developer replenisher. Only the proper amount required should be mixed, and large storage tanks should be avoided. Many 90 second developers have a rather short shelf life, and large amounts should not be prepared in advance.

Transport Cycle versus Phase-in Time

In any given cycle of frames per second of a serial film or cassette changer there are 2 definite and distinct segments. The 1st segment of the cycle is the transportation segment, when the cassette or film is transported into position. The 2nd segment of the cycle is the compression phase, in which the intensifying screens are closed, and maximum compression is achieved for the best possible screen-film contact. During the compression cycle an exposure is made prior to the opening of the cassette, which reinitiates the transport cycle of the next frame. A definite delay occurs in any given cycle after compression is started and prior to the firing of the x-ray unit. There is also a definite delay in the firing of the x-ray machine after the exposure has been initiated, for a series of relays has to close prior to the making of an exposure. This relay delay is not particularly critical in today's electronic systems. Compensation can be made for this problem. There is, however, still a delay prior to exposure even after adjustment of the relay systems. This delay, known as "phase-in time," varies with different types of equipment from 0 to 16 milliseconds. Discussion of this material at any length is beyond the scope of this textbook, being rather the prerogative of the physicist. It is sufficient to mention here that phase-in time can rob the radiographer of useful exposure time per cycle (Fig. 9-5). As higher mA generators become available, phase-in time must be held to a minimum so that exposures in the very short millisecond range can be utilized for angiographic procedures. The technologist should be concerned about the phase-in time of every new piece of generating equipment contemplated for an angiographic room.

Maximum Useful Exposure Length Per Cycle

The part of the feeding cycle during which film is immobile, completely compressed between the intensifying screens differs from the maximum useful exposure length per cycle because of the phase-in time of the x-ray apparatus.

SOME UNIQUE USES FOR SERIAL UNITS

Stereo angiography

Single-plane stereo angiography is carried out with 2 x-ray tubes mounted side by side at a specific angle to each other. A typical stereo angle between the tubes would be approximately 8 degrees. During a stereo run every other film frame is exposed by the alternate tube. When these stereo pairs are viewed in a stereoviewer, stereoscopic pictures result. Early stereo units utilized a single x-ray tube that was moved back and forth mechanically in synchronous action with the exposures and the film transfer mechanism. Newer units utilize electrical switching mechanisms that alternately fire one and then the other tube.

If a stereo run were attempted in which 2 frames per second were used, the result would be 1 stereo pair per second. If a 4 per second run were attempted, the result would be 2 stereo pairs per second.

When a cross-hatch grid is used for stereo angiography in the frontal projection, the grid in the changer should be positioned so that the grid lines run in a conventional manner, permitting conventional angulation technics. The overlay grid (5 to 1 grid) is of such a low ratio that stereo radiographs can be made with no appreciable differences in density. Stereo tubes are usually mounted with 1 tube at a direct right angle to the film, and the 2nd tube in the stereo angle position. There can be approximately a 4 inch difference in the focus-film distance of 1 tube as compared with that of the other. This should not make an appreciable difference in radiographic density. If tube angulation or the increase in focus-film distance results in a density problem, an appropriate tube filter can be used.

Direct Roentgen Enlargement Angiography

If a changer is to be used for direct roentgen enlargement angiography, it should be easily movable and adjustable in height to permit this technic. The grid should be able to be removed quickly with a minimum of effort for the air-gap technic. If the moving of the unit or the removal of the grid takes an appreciable amount of time, there is a tendency to ignore the use of direct roentgen enlargement angiography as a supplementary technic. Additional information on this subject is found in Chapter 4.

Full-Size Film Changers for Conventional Radiography

The modern technologist has achieved tremendous technical latitude, particularly in chest radiography, with the availability of the new collimators in conjunction with high-ratio grids. When one adds a high-speed phototiming mechanism to the collimator and high-ratio grid, high-quality chest radiographs are repeatedly obtained. Patients can be brought directly to the chest radiographic room, examined, and sent immediately back to their rooms with minimal concern for quality control or checking of film prior to the release of the patient. These devices hold 100 or more sheets of 14″ × 17″ radiographic film for routine radiography of the chest (Fig. 9-7). The loading magazine is filled with a box of radiographic film, and photo identification cards similar to the type used for photoroentgenography are used to activate the unit. The cassette is opened, loaded with an x-ray film, and a radiograph is made, all automatically. After the termination of the exposure, the film is emptied into a receiving magazine, and the receiving cassette can be removed for immediate film processing; or it can be left in place to collect dozens of exposed radiographic films. The photo identification card is photographed simultaneously on the film as the exposures are made. Chest films of high quality whether conventional or stereo, can be made easily and seen almost immediately.

The entire chest system can be linked to an automatic processor. The film is taken and automatically fed into a processor, and within 90 seconds a finished radiograph is available. The films can actually be seen before the patient leaves the room, let alone the x-ray department. To some, an operation of this nature might be considered an absolute luxury; yet in a large hospital where hundreds of chest films are taken daily, it becomes almost a necessity. For example, if 200 chest studies are made per day, it means that at least 400 14″ × 17″ cassettes must be handled daily. With an automatic chest unit, at

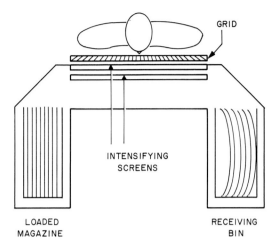

Fig. 9-7. Automatic film changer for routine chest radiography. A film changer designed for quick and easy evaluation of the chest of a large number of patients holds 100 or more sheets of 14″ x 17″ radiographic film in its loading magazine. The unit is activated by the insertion of a photo identification card into a slot on top of the changer, causing a film to enter the cassette. The intensifying screens close, and an exposure is made. The film is then moved into the receiving magazine. This unit can be linked directly to an automatic processor, and the processed chest film can be seen within a few minutes.

least 2 departmental employees, the darkroom technician and the radiologic technologist, are spared the effort of handling these 400 cassettes. A single pair of intensifying screens is used with a system of this type, and can be replaced every few months for the ultimate in intensifying screen output. Careful watch can be kept on the phototimer mechanism to be sure of its giving as quick a response as possible. Keeping 1 chest phototimer constantly fine-tuned is much easier than worrying about a half dozen phototimers dispersed throughout a department. In most of these chest units the collimator is automatically centered to the changer and to the grid, and as the changers move vertically, the collimator tracks automatically. A unit of this type can be installed in a conventional table for routine Bucky radiography.

Angiotomography

A linear body section device with a relatively short exposure angle (10 to 20 degrees) capable of exposures under $\frac{1}{4}$ of a second is commercially available. A "book" cassette (up to 4 pairs of screens per "book") is required to produce multilayered angiotomographic studies. A motor-driven drum holds 4 book cassettes for continuous serial studies. A section is made, using the 1st cassette, the drum is rotated, and the 2nd book is brought into position, exposed, and so on.

REFERENCES

1. Amplatz, K.: New Rapid Roll-Film Changer. Radiology, *90*:130–134, January 1968.
2. Cullinan, J. E., Jr.: The role of the x-ray technician in the cardiopulmonary laboratory. X-ray Techn., *31*:623–627, 631, May 1960.
3. England, I. A.: Angiocardiography: choice of equipment. Focus, *6*(2):4–7, 1965.
4. Feddema, J., Recourt, A., and Monte, G.: Evaluation of image quality in various methods for the recording of movement phenomena. Medicamundi (Eindhoven, Holland), *14*(3):150–154, 1969.
5. Franji, S.: Split field technique using an automatic film changer. Radiol. Tech., *42*(2):80–82, September 1970.
6. Horner, R. W.: Angled bilateral cervical arteriography in the investigation of ischemia. Radiol. Techn., *42*(3):125–132, November 1970.
7. Mattson, O.: Practical photographic problems in radiography with special reference to high voltage technique. Acta Radiol. (Stockholm), Suppl. 120:68, 1965.
8. Rossmann, K., Haus, A. G., and Dobeen, G. D.: Improvement in the image quality of cerebral angiograms. Radiology, *90*(2):361–366, August 1970.
9. Tuddenham, W. J.: Rooms for Film Processing and Display. *In* Scott, W. G.: Planning Guide for Radiologic Installations. ed. 2, p. 68. Baltimore, Williams & Wilkins, 1966.
10. Vennes, C. H., and Watson, J. H.: Patient Care and Special procedures in X-Ray Technology. ed. 2, p. 180. St. Louis, C. V. Mosby, 1964.

10. Basic Electronic Image Detector Principles

For many years fluoroscopists dreamed of brightening the fluoroscopic image. It was hoped that this brightening of the fluoroscopic image could be combined with a reduction in radiation dosage to the patient and recorded by photographic means, like that available in still or motion picture cameras. In the early '40's Chamberlain predicted the development of an electronically enhanced fluoroscope.[6]

IMAGE INTENSIFIERS

Using an image intensifier instead of a conventional fluoroscope definitely shortens fluoroscopic time, for the brighter image is much easier to visualize and interpret, thus facilitating fluoroscopic spot filming. With the intensifier newer vascular technics which require elaborate surgical procedures can be performed in an illuminated room. Elimination of the restrictions placed on the angiographer because of the previously darkened fluoroscopic rooms significantly shortens vascular opacification procedures. Seriously ill, emotionally disturbed, or uncooperative patients can be fluoroscopically evaluated in an efficient and yet concerned manner, for the procedure is more personalized because of full, or partial, room illumination. Another timesaving feature of the electronically enhanced fluoroscope is the elimination of the need to dark adapt the eyes prior to a fluoroscopic procedure. An image intensifier can be equipped with either a mirror-viewing system (Fig. 10-1) or a television viewing system (Fig. 10-2) for quick as well as easy viewing. The mirror is useful when routine procedures are being performed, and the television is invaluable as a teaching aid.

Description of an Image Intensifier

An image intensifier consists of a large vacuum tube containing an input phosphor screen composed of zinc cadmium sulphide crystals. The crystals fluoresce when they are excited by x-ray, forming a light pattern. In proportion to the intensity of the fluoroscopic image, this light pattern is converted into an electronic pattern on a photocathode, which is a thin transparent photoemissive surface in close proximity to the input fluoroscopic screen.[3,7,21] The electrons given off in any segment of the photocathode approximate the amount of light from the same segment of the fluoroscopic image. At the opposite end of the large vacuum tube is a fluoroscopic phosphor smaller than the input phosphor. A typical input phosphor is either 6 inches or 9 inches in diameter, whereas the typical output phosphor is 1 inch or less in diameter. This results in an approximate 6 to 1 or 9 to 1 reduction in fluoroscopic screen size. Although the output phosphor is also made of zinc cadmium sulphide, phosphors of finer grain are used for maximum resolution.

A source of high voltage (usually 25 kVp) is used to accelerate the electrons from the photocathode to the smaller output phosphor screen. A high voltage power supply, coupled with a high voltage cable, is used to cause the electrons to travel at an accelerated speed from the input phosphor to the output phosphor of the intensifier. As the electrons strike the output phosphor with extremely high velocity, there is an increase in the light yield

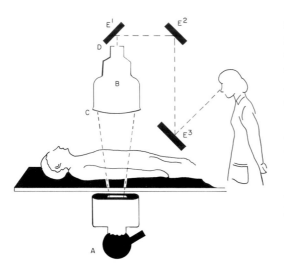

Fig. 10-1. Mirror optics image intensifier. A mirror optical viewing system can be used with an electronically enhanced fluoroscope. *Primary* radiation from the x-ray tube (A) is absorbed or attenuated by the body. *Remnant* radiation strikes the input phosphor (C) of the image intensifier (B). The intensified image at the output phosphor (D) of the intensifier is collected and transferred by a series of lenses and mirrors (E^1 and E^2) to the viewing mirror (E^3).

increased the fluoroscopic screen would fluoresce to a brighter degree, but at the expense of an increase in dosage to the patient. Although a minor increase in kilovoltage somewhat brightens the screen, a significant increase in kilovoltage creates a proportional increase in scattered radiation, and as scattered radiation increases, higher-ratio grids are required.

Although the resolution capability of the viewing systems of intensifiers can be compared with that of the fluoroscopic screen, the viewer sees more details because of the significant increase in light level as compared with the low light level of the normal fluoroscopic screen. Many factors impair resolution in an image intensifier system. Some of the more obvious causes for image unsharpness, or blurring, include geometrical blurring, which can be due to the size of the focal spot, and phosphor graininess, which is caused by the graininess of the individual crystals in the phosphor layers. A major resolution offender is scintillation, otherwise known as x-ray noise or quantum noise.[1] This occurs when lower dosages are utilized on the image intensifier, and very few photons reach the input intensifier screen. The result is quantum fluctuation, which produces a noisy grain-like image.[18]

If one were to utilize a conventional image intensifier with a fractional focal spot (0.3 mm or smaller) for direct roentgen enlargement technic, there is marked improvement in image quality, due as much to the spreading out of a smaller portion of the part under study over the phosphor of the intensifiers as to the fractional focal spot.[9] Unfortunately, the image intensifier is the weakest link in the entire resolution chain. It performs a function similar to that of the intensifying screens of the cassette in routine radiography, but instead of converting radiation to light by a factor of 20× to 30×, the intensifier converts radiation to light by a factor of over 3,000×. As image systems become more complex, the fluoroscopic image can be brightened to a degree of 100,000× or perhaps 1,000,000×, as compared with the

of the unit. The concentration or minification of the light from the smaller output phosphor also has an intensification effect. The electrons do not strike the output phosphor in a random fashion, but are aimed or focused by a system of electronic lenses which electrostatically focus the electron beam so that it is reduced to the size of the smaller output phosphor. When the electron pattern strikes the output screen, it is converted into visible light. Appropriate optical devices are utilized with mirror systems or television systems for viewing (Fig. 10-1 and 10-2).

Resolution difficulties with an Image Intensifier

When conventional fluoroscopy is attempted with typical technical factors, the illumination of the screen is extremely poor, and perceptibility of image details is quite limited. If the milliampere factors were

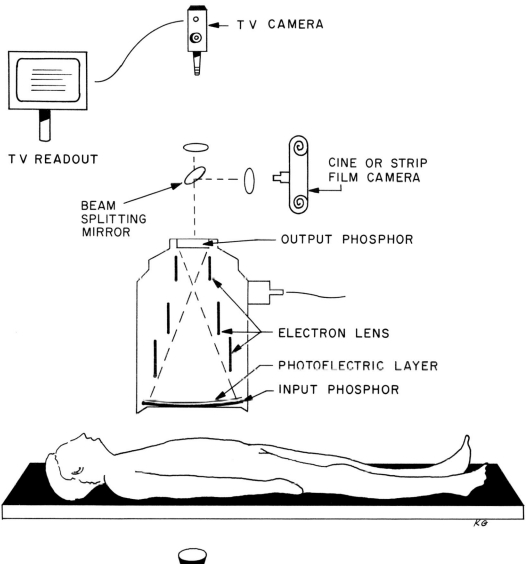

Fig. 10-2. Basic components of an image intensification system. An image intensification device is an electronically enhanced fluoroscopic device. The patient is placed between tube and the input phosphor of the intensifier. *Remnant* radiation as well as *scattered* radiation strikes the input phosphor of the image intensifier.

This light image is converted to an electronic image on the photoelectric layer of the intensifier tube. The electrons are accelerated by the application of a high kilovoltage value to the output phosphor of the intensifier. Electron lenses are used to focus the electron pattern from the photoelectric layer to the output phosphor of the tube. The output phosphor is considerably reduced in size as compared with the input phosphor.

Elaborate beam splitting devices transmit a percentage of the light to the TV camera for fluoroscopic guidance. The remaining light is transmitted to a photographic recording unit such as a motion picture camera or strip film camera (70 or 90 mm). A mirror optical system can be substituted for the television camera.

image of the conventional fluoroscope. This would not be practical, for it would require the use of minute amounts of radiation which would be electronically enhanced to meaningless grainy noisy images.[8]

The intensification factor in the image intensifier is usually expressed as brightness (luminance) gain. Typical commercially available x-ray intensifying systems have brightness gains of 3,000× to 6,000×.

Dual Mode (6 Inch or 9 Inch) Image Intensifiers

Image intensifiers are available with the capability of both a 6 inch input phosphor and a 9 inch input phosphor. The switching mechanism of the dual mode system, electronically controlled, gives the viewer the option of scanning 9 inches of a patient for a large field study, or 6 inches of a patient for a maximum resolution study, such as a cine coronary angiographic procedure. The consensus is that the smaller input phosphors exhibit the greater detail, and many fluoroscopists refuse to use the 9 inch mode or the dual 6 inch or 9 inch mode, preferring the smaller 6 inch input phosphor.

Regardless of the size of the input phosphor utilized, proper primary beam collimation is required to improve the fluoroscopic image as well as the photographic studies. Fluoroscopic shutters must be restricted to input phosphor size, and a proper grid should be employed. In short, every effort must be made to restrict exposure to the smallest possible area of the patient, thereby minimizing the production of scattered radiation.

IMAGE DISTRIBUTORS OR BEAM SPLITTERS

Optical devices known as "image distributors" are used in close proximity to the output phosphor of an image intensifier to provide several channels for optical systems.[4, 14] Many options are available to the fluoroscopist—mirror viewing, television viewing, cine cameras, single-frame photographic cameras (70 or 90 mm strip cameras), and

so on. If some form of image recorder is being used with an image intensification system in conjunction with either a mirror or a TV system, a beam splitter becomes a necessity. These devices reflect a predominant percentage of the light image in one direction, and transmit a smaller percentage in another direction for simultaneous viewing as well as recording technics (Fig. 10-2).

A beam splitter can reflect 90 per cent of the light in one direction, and 10 per cent in another, or perhaps 85 per cent in one direction, and 15 per cent in another, depending on the sytem in use. Some of these image distributors have a distribution of up to 3 ports—1 port can be used for mirror or television viewing, a 2nd for cineradiography, and a 3rd for 70 or 90 mm strip filming technics. Since exposure is significantly increased when photographic systems are utilized, there is a more than adequate level of light to monitor a procedure when an image distributor is used.

Importance of Maintaining a Mirror Viewer. If television viewing is the preference of the fluoroscopist, a provision should be made for substituting a mirror viewer in case of damage to the television system. *A mirror viewer must always be available so that it can easily be substituted for a television system.* Failures in video systems can occur at the most inopportune times, creating a crisis, so that a study which could have been completed in a matter of minutes often must be rescheduled. On occasion, particularly with low light level video systems, it is difficult to examine adequately an exceptionally obese patient. Substitution of the direct-viewer mirror system frequently can salvage a study that otherwise would be of poor quality.

THE VIDEO SYSTEM

Basic Video System

The television camera is combined with a synchronizing generator and television monitor to convert an illuminated image into an electronic signal. This signal is transmitted by either a closed circuit or through the air waves and reconverted as an image for viewing. Several types of television cameras are available; these include the vidicon, the orthicon, and the plumbicon cameras. The vidicon camera tube is relatively inexpensive, but its operation requires rather large amounts of light. The image orthicon camera tube is costly and must be operated at close temperature tolerances. It is extremely sensitive to mechanical shocks, and even minor jolts may cause malfunctions within the system. The major advantage of this tube is that it is extremely sensitive to light and exhibits remarkably good resolution even at lower light levels. The plumbicon camera tube is a type of vidicon camera tube, but it is somewhat more sensitive to light stimulation than the vidicon tube and is said to have better resolution than the vidicon tube.[20]

Although vidicon camera tubes are relatively inexpensive, and their low sensitivity to light can be overcome by the increased gain intensifiers now available, they nevertheless exhibit a "stickiness" or "lag." This persistence of the fluoroscopic image on a television screen can be annoying when areas of high detail are being examined.

The camera tube, which is the eye of the system, converts the 150,000 elements that usually make up an average picture into an electronic image by the use of a scanning beam within the tube.[20] The beams scan diagonally downward on the screen from left to right, reproducing the image picked up by the camera tube. Conventional television systems utilize a 525 line scan or line pattern.

Television systems are commercially available which use 625 line patterns or 875 line patterns, and there is a general feeling that greater image sharpness and resolution are attained with an increase in the line pattern capability of a television system. In effect, the 875 line pattern should be sharper than the 525 line pattern.

All television systems are quite helpful in improving fluoroscopic contrast since the output phosphor of an image amplifier exhibits extremely low contrast.

Advantages and Disadvantages of a Television System

The utilization of a television system coupled to an image intensifier has major advantages. Since more than one person can view the television image, this system makes an ideal teaching tool. Electronic enhancement of the television image also increases image details. Image contrast can be improved by the television screen as well as by a video tape recorder or a video disc scanner. Radiation levels are definitely decreased when television recording methods (disc scanners or tape recorders) are used, as contrasted with the level required in cineradiographic technics, for the low radiation factors required for fluoroscopy can be used with the video recording systems. A significant disadvantage of the television system is the scanning line type of picture, which can be interrupted by electronic noise.

Video Tape Recorders. The use of video tape recording has proved quite helpful to hospitals in many areas, including surgery, nursing, and all forms of teaching. The price range of the video tape recorder has been materially reduced with the availability of transistorized television tape recorders. The tape recorder makes it possible for a technologist to fluoroscope without the aid of a radiologist, who can view the televised fluoroscopy at a later time. Video tape recording systems can also be used as an audio-visual aid for teaching student radiologic technologists.

A comparison of video tape recordings with cineradiographic recordings will show that video tape has various advantages. The lower radiation dose is the major advantage, but the availability of instant playback without processing and reutilization of the tapes must be emphasized. There are disadvantages, of course, such as the poor resolution capability of the television system, as opposed to a photographic system. The slow-motion technics available with video tape recorders are not even close to the quality of the slow-motion effect of cineradiographic film, with which studies are performed at a rate of 60 frames per second or greater.

When a video tape recorder is utilized with a video display system in conjunction with an image intensifier, the fluoroscopic image on the output phosphor can be displayed directly on a television monitor, and simultaneously recorded on magnetic tape or a video disc scanner.[9] A major advantage of the tape recorder is that the study can be seen immediately without processing and can be viewed over and over. An angiographer attempting a complex procedure could devote his full time to the procedure while simultaneously taping the procedure, and immediately view the tape when the pressure of the moment has been lifted. Although the video tape recording has a slightly lower resolution than the original television picture, it is of very high quality. The major limitation of the video tape system is that it has to be played back on a video tape recorder. These recorders are not as available as film projectors. Cineradiographic films can be shown on many types of projectors. These projectors are generally available, and can be obtained when cine presentations are being given outside of one's department. This is not true of video tape recordings.

Video Tape Recording of Angiographic Studies. The use of a video tape recorder linked with an intensifier in combination with a see-through serial film changer can be quite advantageous to an angiographer. The success of an injection can be determined before the roll or cut film series has been processed. When small, carefully controlled injections of opaque are used to ascertain the positioning of a catheter, the injection can be made, the tape reversed, and the situation evaluated by the angiographer.

Positive Recall Video Tape Systems. Until recently the finding of a particular sequence on video tape, which can be hundreds of feet in length, was at best cumbersome. New systems are available using electronic keying devices that enable a fluoroscopist to find immediately a specific sequence, regardless of its position on a lengthy video tape. A master video tape recorder can be kept in a remote area for use with multiple monitors strategically located throughout the radiographic facilities of the hospital. The fluoroscopic work of an entire day can be stored on the master tape recorder. When a referring physician requests a consultation, not only can the films be shown to the physician, but the video tape fluoroscopy can be recalled from a remote location if a positive recall video tape system is available.

Video Disc Scanner. As with all complex electronic equipment, the video disc scanner is somewhat expensive, but there is no question that its cost will soon decrease. In the late '50's a video tape recorder cost over $100,000, and now most commercial video tape recorders for use with an image intensifier can be purchased for under $10,000. Home recording video tape recorders of good quality are commercially available in the $1,000 range. It is only a matter of time until the video disc scanner becomes commercially available in a reasonable price range.

A scanner is capable of freezing a single television frame, reducing dosage to both patient and operator.

Video disc recorders can record and play back video images as normal, slow, reverse, or stop-motion studies, and it is a simple matter to locate an individual frame rapidly. This is a versatile way of storing thousands of individual video frames for immediate or delayed analysis. A conventional video frame rate is 30 frames per second. Video disc scanners using dual disc systems can record over 2,400 frames in a period of 80 seconds. Video disc scanners utilize magnetic particles deposited on a rigid disc rather than a flexible tape. The disc revolves at high speed (approximately 1,800 rpm) on a turntable.

The major advantage of a disc recorder over the video tape recorder is that when an individual frame is viewed, there is no wearing away of the image. Tape recorders, when stopped, produce a wearing of the tape, and if there is continual viewing of a single area on a video tape, damage can eventually be caused to the tape, resulting in loss of the picture. This is not true of the video disc recorder.

Rich compares the video tape to the cine camera, and the video disc scanner to the fluoroscopic spot film, since each unit is intended for a different purpose. One should no more compare the video tape recorder to the video disc scanner than compare the cine camera to a spot film system.[19]

Image storage systems such as the video disc scanner "freeze" an x-ray image, providing image comparisons and subtraction technics. The systems have a stop-action storage feature, lowering radiation dosage to the patient. Short bursts of radiation, approximately $\frac{1}{10}$ of a second long, are used to create the "freeze" effect. The physician can then study the stored image on the television screen without further dosage to the patient. Spot films can be taken without the use of radiographic film, to be evaluated at a later time. A short burst of x-ray could be used to "freeze" a projection of the hip in the AP position during a hip pinning. This spot film of a sort could be evaluated on a television monitor by the surgeon, the guide wire or pin could then be readjusted, and a 2nd pulsed spot film taken, using the "freeze" system, the image on the storage tube would be evaluated, and, if necessary, the initial projection could be called back for comparison. One projection could be superimposed on the other on the screen to see if there had been a shifting of the metallic orthopedic device or of the fragments of the injured femur.

Pulsing devices are available with these systems that expose the patient to a small amount of radiation for an extended period of time. For example, an exposure can be made ($\frac{1}{10}$ of a second) and then repeated every second for a "lag" effect. This lowers the dosage to the patient and to the operator, and a stop-motion effect is created. When these images are viewed in sequence, definite physiologic occurrences can be seen but without the smoothness of a cineradiographic or video tape study.

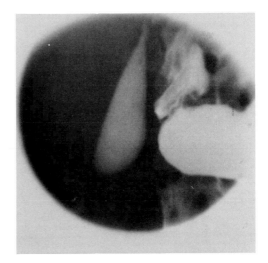

Fig. 10-3. A 70 mm kinescopic spot film taken from a 1,000 line television monitor exhibits extremely high resolution of the opacified gallbladder.

Kinescopic System. Kinescopic recording involves the photographing of an image from a television screen. It can be described as still or motion picture photography of the monitor of a closed circuit television system[19] (Fig. 10-3). The quality of the kinescopic image is limited to the quality of the image on the television monitor, and a definite disadvantage is that there is no possibility of rapid playback. Since television monitors generally display 30 frames per second, it is difficult to create a slow-motion effect with a kinescopic study. Elaborate electronic systems do permit kine recordings at 60 frames per second. Unfortunately, the motion picture camera records all of the noise inherent in the television system. A significant bonus is that lower patient dosages are received in the use of the kinescopic system. It is possible to kinescope a televised image from a video tape recording at a later date. Video tape recordings could be made of all fluoroscopic procedures and edited at the convenience of the fluoroscopist. A motion picture camera could be aligned with the television monitor that is used to view the video tape recording, and if a specific fluoroscopic sequence required a permanent recording, a kinescopic recording could be made by the viewer.

CINERADIOGRAPHY

Cineradiography is a means of preserving the fluoroscopic image. If a physiologic event is too fleeting or too rapid for normal fluoroscopic viewing, it lends itself well to cineradiography. Cineradiographic units are now relatively easy to operate and produce cineradiographic films of excellent quality; the radiation to the patient, considering the information gained, is minimal. The input field sizes of available image intensifiers vary from 5 inches to 9 inches. A popular image amplifier used for cineradiography is a dual mode system. Six inches of the input phosphor are utilized for high-detailed cine work, and the entire 9 inches of the input phosphor for larger body areas.

Two Methods

There are 2 methods of cinematography—the direct and the indirect method.[14, 24]

The direct method consists of exposing photographic films in direct contact with intensifying screens. This method produces an image of excellent quality that is at least as large as life size, but at the sacrifice of minimal frames per second. The most elaborate commercial serial film changer is limited to a rate of 12 films per second. Serial film or cassette changers as we know them today, are inadequate for cine studies. Even if a serial film changer could produce 24 radiographs per second, and if each individual frame were motion-picture-recorded as a life-size cinematographic run, we still would not have the slow-motion capability of present-day image-intensifier-cine recording systems.

The indirect method of cineradiography includes all types of image gathering systems, whether solid state, electron optical, or mirror optical units in combination with an image intensifier. The most popular method of cineradiography is the recording of the image from the output phosphor of an image intensifier tube. There is definite unsharpness due to the graininess of the output phosphor as well as to the grain of the faster films required for high filming rates. The present section of this textbook is confined primarily to cineradiography of the image of an output phosphor of an image intensifier tube.

Cineradiographic Film Exposure Methods

Three major methods are available for exposing cineradiographic film in conjunction with an image intensifier.

1. *The Continuous X-ray Exposure Method.* A continuous beam of radiation forms a continuous image on a fluoroscopic screen. This is not desirable from the standpoint of exposure level, for most of the radiation does not contribute to film density. For example, if a 10 second cine run is attempted, the patient is exposed to 10 full seconds of radiation. During this 10 second period the cine camera opens its shutter, records an image from the output phosphor, closes its shutter, permitting the film to advance for the next frame, opens its shutter again, and so on. The cine camera records only half of the exposure that the patient receives (Fig. 10-4A). Not only does the patient receive more dosage than that required to expose the film, but the continuous exposure is traumatic to the x-ray equipment, particularly the x-ray tube. This type of cine exposure system is known as a *nonsynchronized system.*

2. *The Synchronized System.* In this system radiation exposure is on when the cine camera shutter is open and off when the cine camera shutter is closed (Fig. 10-4B). This is a form of pulsing of the image accomplished at the primary of the high-tension transformer. The synchronous pulsing of the x-ray exposure significantly lowers the radiation dosage to the patient as compared with the exposure from the continuous x-ray or nonsynchronized system.

3. *The Grid Pulse System.* Grid-control x-ray tubes are manufactured with a 3rd electrode in the cathode assembly.[16] The 3rd electrode, called a "control grid," is used to control the flow of electrons in an x-ray tube.

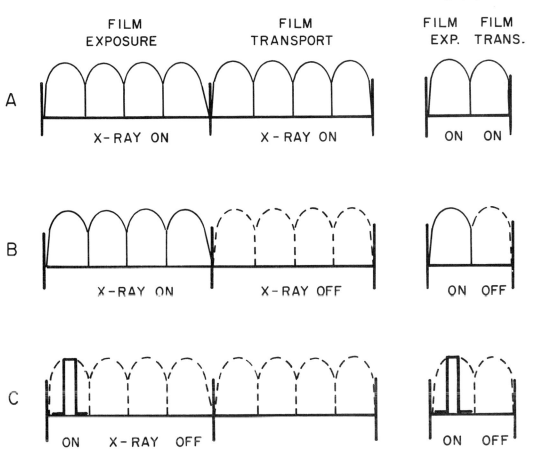

Fig. 10-4. Cine film exposure methods. Three major systems are used to expose cineradiographic film in conjunction with an image intensifier.

(A) *Continuous x-ray exposure method.* Radiation exposure is on continually through both the film exposure and the film transport stage. The patient receives significantly more radiation than what is required to expose the film. This method is known as the *nonsynchronized system.*

(B) *Synchronized system.* Radiation exposure is on only when the camera shutter is open. Less radiation is given to the patient in this method than in the nonsynchronized method (A).

(C) *Grid pulse system.* A short burst of x-ray (in the millisecond range) is used with this method only for a portion of the time that the camera shutter is open. The tube operates only when the cine film is standing still, and kinetic blurring of each individual frame is avoided or at least decreased as compared with systems A and B. Each frame is photographically sharper, and dosage is held to the required minimum.

It permits extremely accurate timing with very short exposures. The cathode head is especially wired so that it can act as a grid. A "bias" voltage is applied to the cathode head. This bias voltage does not permit the continual flow of electrons across the x-ray tube to the anode, although the filament is heated for maximum emission. When the voltage level drops to zero, the preset tube current begins to flow across the tube immediately, making it possible to synchronize x-ray exposure with the shutter of a motion picture camera (Fig. 10-4C). Current flowing through the tube can be stopped, even though high-voltage values are impressed upon the tube. This off-and-on repetitive process is called "pulsing" of the tube. The radiation pulse can be in the range of 2 to 4 milliseconds, or as quick as 1 millisecond ($\frac{1}{1000}$ of a second) during the shutter-open stationary phase. The tube fires only when the cineradiographic film is standing still, and any kinetic blurring of the individual film frame is thus avoided, with an important reduction in radiation to the patient, as opposed to the nonpulsed system.

Comparison of the 3 Methods. Let us now compare all 3 systems: the continuous exposure method, the synchronized pulse and the grid pulse system. If a cine camera were to operate at 60 frames per second, and the open shutter time were $\frac{1}{120}$ of a second, then the closed shutter time, including film transportation, would be $\frac{1}{120}$ of a second. With the continuous exposure technic, the patient would receive radiation for the entire cine run, or exposure for 1 full second of each 1 second of the cine study. With the on-off system of the synchronous pulse technic, the patient would be exposed on alternate periods for $\frac{1}{120}$ of a second, for a total of one half second exposure for each 1 second of cine run. With a cine pulse technic utilizing a grid-controlled tube and having exposure capabilities of 2 milliseconds per frame, the patient would be exposed to 60 frames × 2 milliseconds, or 120 milliseconds per second. This represents an exposure to the patient of less than $\frac{1}{8}$ of a second (Fig. 10-4). Individual frames of a pulsed run can be projected and viewed as single frames with much more sharpness of the image, resulting in more diagnostic information. Many of the cine pulse cameras are capable of 80 frames per second, with high milliampere output capabilities.

Automatic brightness stabilizers can be used with these systems. Automatic bright-

ness is a form of phototiming, automatically increasing either kilovoltage or milliamperage when thicker body parts are encountered. From the standpoint of scattered radiation, variation of milliamperage can have a significant advantage in studies of thicker patients, as compared with a kilovoltage compensation system. Most automatic brightness control systems give the operator of the unit the option of manual control. Whether a manual or an automatic brightener system is utilized, it serves the same purpose—to obtain during cineradiography, as patient thickness or opacity varies, optimal brightness of the image intensifier as well as good contrast.

A definite disadvantage of any cine technic is that cinefluorography requires at least 10 times more intensity than the energy value used in fluoroscopy.[23]

Cine Frame Speeds

Most cineradiographic cameras can function as slow as 8 frames per second up to the more conventional 64 frames per second. Some of the newer 35 mm cameras have an 80 frame per second capability, whereas some 16 mm cameras are available with a 200 frame per second capability. A leading manufacturer offers frame rates of $7\frac{1}{2}$ frames, 15 frames, 30 frames, or 60 frames per second.

It is interesting to multiply the frame rate per second of a camera by the actual length of a cine run, to ascertain the total number of individual frames that a radiologist is required to interpret.[13] For example, if a 5 second cine run were attempted with an 80 frame per second camera, 400 radiographic images would be available for evaluation. Despite this great number of individual miniature radiographs, a diagnosis is frequently made from only a few of these frames. Since there is very little time lost between individual frames at high speeds, a rapidly moving vessel or organ can be carefully evaluated.

A high-speed experimental camera is available with continuous noninterrupted film transport (540 frames per second); using a rotating prismatic shutter, 100 feet of 16 mm film can be exposed in $7\frac{1}{2}$ seconds (at 540 frames per second) or in 15 seconds (at 270 frames per second). The authors claim that the exposure time per frame is 1/2700 of a second. Experimental cine studies are being performed on small dogs or piglets.[22]

Cine Frames per Second. Cineradiographic procedures can be compared with conventional motion picture studies on a basis of the number of frames per second. Conventional motion picture film with sound tracks requires 24 frames per second. Silent motion pictures can be made at 18 frames per second. When slow-motion films are desired, the frame per second rate of the camera is speeded up to record more frames per second than the technologist intends to project. If 24 frames were to be made and then projected at 24 frames per second, the viewer would have the impression of a moving image. If 60 frames per second were to be made and projected at 24 frames per second, it would take $2\frac{1}{2}$ times longer to view the finished motion picture study, and the result is slow motion. Conversely, if one wished to speed up a motion picture study, it could be made at 12 frames per second, and shown at 24 frames per second. Events would seem to take only half the time in this speeded up film.

An increased camera rate of frames per second, combined with a conventional projection rate, can be used to advantage in evaluating swallowing function, cardiac pulsations, or any arteriographic procedure. The minimal frame recording technic combined with a conventional projection technic lends itself well to areas where motion is somewhat limited or sluggish—for example, the functioning of the stomach with a slow emptying time.

Biplane Cineradiography

Before biplane cineradiography can be performed to best advantage, there should be 2 grid pulse systems. The pulses can be triggered alternately. A frame is made in the frontal position while there is a blankout field in the lateral position, and vice versa. Scattered radiation from one plane does not damage the image from the opposing plane.[10]

Types of Film for Cineradiography

Both 16 mm motion picture film as well as 35 mm motion picture film are in common use for cineradiographic procedures. The major difference between 16 mm film (40 frames per foot) and 35 mm film (16 frames per foot) is that 35 mm film has approximately 4 times more silver halide grains to be exposed than 16 mm film, for each frame is approximately 4 times larger. The dimensions of each frame of a 16 mm film are 10.5 mm by 7.5 mm, whereas 35 mm film has a frame size of 20 mm by 18 mm[3] (Fig. 10-5). Therefore, for every individual grain of silver halide available on a 16 mm film to record a minute image detail, there are 4 or more grains of silver halide on a 35 mm film.[15] The amount of x-ray energy required to blacken 16 mm film is not adequate for exposing 35 mm film, which requires from 2 to 4 times more radiation to achieve the same degree of blackness.[17]

Accessories and Evaluation. Cineradiographic accessories, such as film splicers, film editing equipment, film viewing equipment, and so on, are more readily available in the 16 mm size. Although devices of this nature are available in the 35 mm size, they are considerably more expensive than their 16 mm counterparts. The storage of completed cineradiographic studies presents other new problems, which are compounded by the use of 35 mm film, for this requires $2\frac{1}{2}$ times the length of film, due to its increased frame size, of a comparable 16 mm study. Not only is the film longer, but it is twice as wide, taking up more storage space. There is a general ease in handling 16 mm film, it is considerably less expensive, and there is less dosage to the patient. In the final analysis, film size, as well as film type, is left to the cinefluoroscopist, who must determine his own needs. The selection of such film by

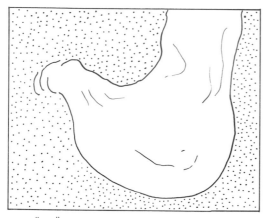

4" x 5" SPOT RADIOGRAPH (4 ON I , 8"x I0")

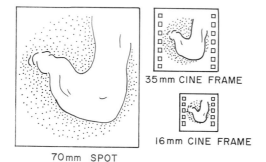

35 mm CINE FRAME

16mm CINE FRAME

70mm SPOT

Fig. 10-5. Frame size comparison. Four different film frame sizes are relatively approximated. A 4″ x 5″ segment of a routine 4 exposures on 1 (8″ x 10″) fluoroscopic spot film is compared with a series of photographic film images made from the output phosphor of an image intensifier. Shown are a 70 mm spot film, a 35 mm and a 16 mm cineradiographic frame.

As frame size increases, radiographic detail should increase.

make and relative speed will not be discussed in this textbook.

FURTHER COMPARISONS OF IMAGE RECORDING

Comparison of Motion Recording Technics

No one method of image recording, whether it be a conventional radiograph or a 540 frame per second cineradiographic study, is the final answer. If dynamic functional studies are required, then motion picture studies or video tape recordings are helpful, but the best recorded detail is still seen on conventional radiographs. Multiple methods of recording must be available to the radiologist. From the standpoint of motion picture studies, cineradiographic studies directly from an output phosphor exhibit the highest radiographic details. A video tape recorder can be substituted for some cineradiographic procedures. A kinescopic examination can be used if reduced radiation exposure is a consideration. Although video tape is more practical, cine is more permanent. If a slow-motion effect is desired, cinefluorography, with its rapid frames per second, produces the most realistic slow-motion effect without a flickering look.[19] A video tape recorder must be considered if economy is a factor, for the tape can be erased and reused any number of times. The obvious advantage of the video tape recorder is that no photographic processing is required. The disadvantages of the cine and kine systems are the processing of large amounts of film and the need for subsequent editing.

Salter claims that a television scanning system, utilizing 525 lines per picture frame, results in about the same amount of detail as that available on a 16 mm motion picture film. He claims that a single frame of 35 mm motion picture film has about 500,000 picture elements, whereas a frame of 16 mm film has about 125,000 picture elements. Commercial television in the United States, transmitting at a rate of 525 lines per picture frame, allows for the transmission of about 150,000 picture elements per frame.[20] Since the image quality of a video tape recorder is almost a duplicate of the initial video image, it follows that the video tape image should exhibit approximately the same detail as a 16 mm motion picture study. Video tape slow-motion technics unfortunately are not flicker-free; therefore, when slow-motion studies are required, high-speed cineradiographic runs must be resorted to, even at the expense of increased radiation dosage to the patient.

Cineradiography versus Serial Film or Cassette Changers

Motion recording systems supplement full-size individual radiographs. A small, almost unnoticed change on a conventional radiograph can be more obvious on a motion picture run. If a comparison is made of the frame speed of a motion picture camera and the frame speed of a serial film changer, for example—60 frames per second versus 6 films per second over any given time period— there will be 10 individual cine frames for each serial film, although the individual frames of the cine are of poor quality as compared with those of the serial film angiogram. More important is the time interval between frames. For a 60 frames per second run, using $1/120$ of a second exposure per frame and $1/120$ of a second for shutter closing and film transportation, the images (60 per second) are separated by $1/120$ of a second time intervals. In other words, an exposure is made, and $1/120$ of a second later another exposure is made. But a serial film changer, running at 6 films per second and using $1/120$ of a second as an exposure time, has serious time gaps between exposures. For example, an exposure is made at $1/120$ of a second, the film changer opens, the film is discharged into a loading bin and a new film replaces it, screen contact is reapplied, and a 2nd exposure is made. There is a time delay of $19/120$ of a second between exposures, 19 times longer between frames than in the cineradiographic studies.

Definite disadvantages characterize the cine recording systems as compared with the film changer systems. The most significant is less detail per individual frame, for motion picture film in use with an image intensifier has a tendency to exhibit more grain. Restricting the cine size to the 6 or 9 inch input phosphor, as opposed to 14 × 14 inch serial film changers, can also be a major disadvantage. Cine processing and editing technics

are cumbersome and time-consuming. It is quite frustrating to find that frame after frame of a study is of no value to the viewer. Special cine projectors with variable speed projection capabilities are required. Frame by frame projection, as well as slow-motion features are necessary, and the projection device must be capable of projecting or viewing a single frame of a film for an unlimited amount of time without damage to the cineradiographic film.

The undisputed advantage of serial radiography with full-size film changers is better image quality, since the serial image is about 1500× larger than that of a cineradiographic frame.[5]

70 MM AND 90 MM STRIP FILM CAMERAS

Rapid serial 70 mm and 90 mm spot films can be taken from the output phosphor of an image intensifier[25] (Fig. 10-5). These technics require only about $1/20$ the exposure needed for a conventional full-size radiograph when a 9 inch input phosphor field is used. With the use of a 5 inch input phosphor intensifier, the factor will be about $1/6$ of the exposure required for conventional film.[11] Geometric and electronic enlargement technics can produce results comparable to those in actual-size radiographs. In the electronic enlargement technic only about $1/5$ to $1/8$ of the exposure required for the full-size radiograph is necessary for 70 mm film.[2] Yet almost the same detail is perceptible as in the full-size radiograph, and there is a considerable reduction to the patient in radiation dosage, as well as a saving in film cost. Examination time is shortened, since the spot film device does not have to be loaded and reloaded with cassettes. Strip film cameras can be operated at 1, 3, or 6 frames per second, and many of these units hold a 400-frame supply. It is important if a strip film camera is being contemplated to make sure of the ease of removal of the take-up cassette. A take-up cassette with a self-

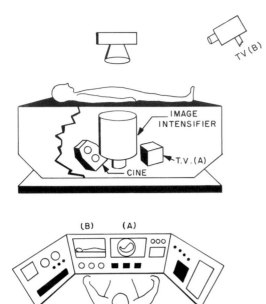

Fig. 10-6. Remote control fluoroscopy. A typical remote control table consists of an image intensifier with a television camera (A) as well as a cineradiographic camera. Many remote control consoles are housed in the room with the x-ray table. In the near future the operating panel could be mounted at a truly remote distance, several rooms, several miles, or even several hundred miles distant. A radiologist or radiologic assistant sits at the control console and has direct sound communication with the patient. A television camera in the radiographic room (*top*, B) monitors the patient and can be viewed on the control console (*bottom*, B). The fluoroscope image is viewed on monitor A (*bottom*).

Remote control radiography and fluoroscopy will enable a fluoroscopist to function in several places at one time from a single operating point.

threading feature makes it possible to process individual studies without waiting until an entire roll of film has been exposed.

REMOTE CONTROL RADIOGRAPHIC UNITS

Remote control fluoroscopic units have been in use for over 10 years and are finally increasing in popularity.[12] A routine remote control installation would include a table with an image intensifier (Fig. 10-6, *top*). A remote control booth separated from the table by 6 to 10 feet would house a control panel (Fig. 10-6, *bottom*). A television monitor displaying the image from the output phosphor of the image intensifier is situated in the control booth, and an audio (sound) system is used for 2-way communication between the patient and the fluoroscopist. An additional video (sight) system is sometimes incorporated within the control panel so that the fluoroscopist can simultaneously observe the patient on one television system and the fluoroscopic image on another (A and B in Fig. 10-6, *bottom*). This additional video system can be helpful if the remote control booth is located at a considerable distance from the table. A remote control setup of this sort could be used miles from a patient, as opposed to a few feet. Radiologic assistants could be used for conventional fluoroscopic procedures, and only when a unique fluoroscopic problem arose would a radiologist be consulted by remote control. The radiologist could have in his film-viewing area a control panel that could operate a number of remote control tables.

THE USE OF THE IMAGE INTENSIFIER BY THE RADIOLOGIC TECHNOLOGIST

Routine use of an image intensifier as a positioning tool is predicted for the radiologic technologist (Fig. 10-6). By using a minimal amount of radiation both centering and positioning could be guaranteed. Phototiming systems would function more efficiently if proper centering of the part to the phototiming pickup device could be guaranteed. Difficult positions could be centered and correctly angled under fluoroscopic guidance. Poor patient preparation could be spotted immediately before attempting an opaque injection. When a fracture is suspected, fluoroscopic scanning would determine the extent of an injury, and a minimal number of films would be taken, thus sparing the patient additional trauma. The role of the radiologic technologist will be significantly expanded as image detection devices become increasingly complex. It behooves the interested and concerned radiologic technologist to learn early and as much as possible about each new technical or electronic accomplishment.

REFERENCES

1. Albrecht, C., and Oosterkamp, W. J.: The evaluation of x-ray image-forming systems. Medicamundi (Eindhoven, Holland), 8(6):106–115, 1962.
2. Becking, H. B., and Hupscher, D. N.: 70 mm image intensifier fluorography. Medicamundi (Eindhoven, Holland), 12(3):95–102, 1968.
3. Bloom, W. L., Jr.: Image Intensification and Recording Principles. New York, General Electric, 1963.
4. Botden, P. J. M.: New developments in image intensification and x-ray television. Medicamundi, 5(2/3):52–60, 1957.
5. Bruns, H. A., and Keck, E. W.: Cineangiocardiography in pediatric cardiology. Medicamundi, 15(1):13–17, 1970.
6. Chamberlain, W. E.: Fluoroscopes and fluoroscopy. Radiology, 38:383–412, April 1942.
7. Coltman, J. W.: Fluoroscopic image brighting by electronic means. Radiology, 51:359–367, September 1948.
8. Combee, B., Botden, P. J. M., and Kühl, W.: Progress in image intensification. Medicamundi, 8(6):101–105, 1962.
9. Cullinan, J. E., Jr.: The role of the x-ray technician in the cardiopulmonary laboratory. X-ray Techn., 31:623–627 and 631, May 1960.
10. England, I. A.: Angiocardiography: choice of equipment. Focus, 6(2):4–7, 1965.
11. Feddema, J.: Consideration on x-ray examinations carried out by remote control. Medicamundi, 14(2):74–84, 1969.
12. Jutras, A.: Teleroentgen diagnosis by means of video-tape recording (editorial). Amer. J. Roentgen., 82:1099, 1959.
13. Knudtson, O. C.: A study of the cinefluorographic aspect of image amplification. X-ray Techn., 33(1):9–17, July 1961.
14. Lewis, R. J.: Television x-ray recording systems. Radiography, XXXVI(422):29–34, February 1970.
15. Lignon, A., and Botden, P.: Image intensification cinefluorography. In Ramsey, G. H. S., Watson, J. S., Jr., et al. (eds): Proceedings of the First Annual Symposium on Cinefluorography. pp. 11–23. Springfield, Ill., Charles C Thomas, 1959.
16. Novak, S. E.: Space age advances in radiology. Radiol. Techn., 36(1):6–13, 1964.
17. Olden, R. A.: Cinefluorography. X-ray Tech., 33(3):163–177, November 1961.
18. Oosterkamp, W. J.: Progress in recording the radiographic image. Medicamundi, 11(4):118–123, 1966.
19. Rich, J. E.: The Theory and Clinical Application of Television in the Radiology Department. New York, General Electric, 1969.
20. Salter, P.: Television equipment in radiography. The Focal Spot, 27(128):143–150, May 1970.
21. Stanton, L.: Basic Medical Radiation Physics. p. 283. New York, Meredith Corporation, Appleton, 1969.
22. Stauffer, H. M., et al.: High speed (540 frames/sec.) bi-plane cineradiography. X-Ray Bulletin, No. 11:14, 1968.
23. Ter-Pogossian, M. M.: The Physical Aspects of Diagnostic Radiology. p. 356. New York, Harper and Row, 1967.
24. Tristan, T. A.: Methods of cinefluorography. Med. Radiogr. Photogr., 35(2):38–44, 1959.
25. Wohl, G. T., and Koehler, P. R.: Experience with 70 mm camera in Image Amplification Fluorography. Clin. Radiol., XVI(4):363–368, October 1965.

Index